This book is composed of a set of chapters contributed by past and present collaborators of the Nobel Laureate Sir Andrew Huxley covering the areas of muscle research to which he has made major contributions. The purpose of the book is to discuss the way that muscle works, asking questions at a functional level about the molecular basis of muscle tone production and muscle contraction. The majority of the chapters are concerned with muscle physiology and the relation between structure and function. The process of activation of muscle is examined, together with the mechanism of contraction itself. Although most of the book is concerned with vertebrate skeletal muscle, several of the chapters deal with cardiac muscle. The book also features two introductory chapters discussing Sir Andrew's achievements in both nerve and muscle physiology.

All those interested in the structure and function of muscle, or cell motility in general, will wish to read this book.

Muscular Contraction

Muscular Contraction

Edited by
Robert M. Simmons
Division of Biomolecular Sciences
King's College London

CAMBRIDGE
UNIVERSITY PRESS

CAMBRIDGE
UNIVERSITY PRESS

32 Avenue of the Americas, New York NY 10013-2473, USA

Cambridge University Press is part of the University of Cambridge.

It furthers the University's mission by disseminating knowledge in the pursuit of education, learning and research at the highest international levels of excellence.

www.cambridge.org
Information on this title: www.cambridge.org/9780521417747

© Cambridge University Press 1992

First published 1992

A catalogue record for this publication is available from the British Library

Library of Congress Cataloguing in Publication data
Muscular contraction / edited by Robert M. Simmons.
p. cm.
Consists of up-dated papers originally presented at a conference held in association with a meeting of the Physiological Society in Cambridge, England in June 1989.
ISBN 0-521-41774-0 (hardback)
1. Muscle contraction – Congresses.   I. Simmons, Robert M. (Robert Malcolm), 1938–   . II. Physiological Society (Great Britain)
[DNLM: 1. Muscle Contraction – congresses.   2. Muscles – chemistry – congresses.   3. Muscles – cytology – congresses.   4. Muscles – physiology – congresses.   5. Myocardial Contraction – congresses.
WE 500 M9873 1992]
QP321.M896   1992
596´.01852 – dc20
DNLM/DLC
for Library of Congress   91-34083

ISBN 978-0-521-41774-7 Hardback

Contents

A volume in honour of
Sir Andrew Huxley, O.M., F.R.S.

Contributors

Lucy M. Brown. Department of Physiology, University College London, Gower Street, London WC1E 6BT, England.

Makoto Endo. Department of Pharmacology, Faculty of Medicine, University of Tokyo, Hongo, Bunkyo-ku, Tokyo 113, Japan.

Lincoln E. Ford. Cardiology Section, Department of Medicine, University of Chicago, 5841 S. Maryland Avenue, Chicago, IL 60637, U.S.A.

Clara Franzini-Armstrong. Department of Anatomy, University of Pennsylvania, Philadelphia, PA 19104-6058, U.S.A.

Yale E. Goldman. Department of Physiology and Pennsylvania Muscle Institute, University of Pennsylvania, School of Medicine, Philadelphia, PA 19104-6085, U.S.A.

Hugo González-Serratos and Maria del Carmen Garcia. Department of Biophysics, University of Maryland, School of Medicine, Baltimore, MD 21201, U.S.A.

Albert Gordon. Department of Physiology and Biophysics, SJ-40, University of Washington, School of Medicine, Seattle, WA 98195, U.S.A.

Jan Lännergren. Department of Physiology II, Karolinska Institute, S104 01 Stockholm, Sweden.

Vincenzo Lombardi and Gabriella Piazzesi. Dipartimento di Scienze Fisiologiche, Università degli Studi di Firenze, Viale G. B. Morgagni 63, 50134 Florence, Italy.

Rolf Niedergerke and Sally Page. Department of Physiology, University College London, Gower Street, London WC1E 6BT, England.

Kenneth Roos. Cardiovascular Research Laboratory, Department of Physiology, UCLA, School of Medicine, 10833 LeConte Avenue, Los Angeles, CA 90024-1760, U.S.A.

Reinhardt Rüdel and Bernd Fakler. Department of General Physiology, University of Ulm, Albert-Einstein-Allee 11, D-7900 Ulm, Germany.

Robert M. Simmons. Biophysics Section and MRC Muscle & Cell Motility Unit, Division of Biomolecular Sciences, King's College London, 26–29 Drury Lane, London WC2B 5RL, England.

Robert Stämpfli, Websweilerstrasse 34, 6650 Homburg 9, Germany.

Stuart Taylor. Department of Pharmacology, Mayo Foundation, Rochester, MN 55905, U.S.A.

Saul Winegrad. Department of Physiology, School of Medicine, University of Pennsylvania, Philadelphia, PA 19104-6085, U.S.A.

Preface

Andrew Fielding Huxley was awarded the Nobel Prize for Physiology or Medicine in 1963, together with A. L. Hodgkin and J. C. Eccles, for his contribution to understanding the process of nerve conduction. But by this time his research efforts had been redirected towards muscle contraction for some twelve years, and he had already played a major role in formulating the sliding-filament theory and in elucidating the first steps in the activation process. During these years and subsequently, he worked with a number of visitors to his laboratory, chiefly postdoctoral fellows from abroad. The opportunity for these collaborators to pay tribute to his influence on their work came at a conference associated with the meeting of the Physiological Society in Cambridge in June 1989. This marked no particular anniversary, though it preceded his retirement from his post as Master of Trinity College by a year. I asked him what kind of a meeting he would like, and he replied without hesitation, "On muscle." So while this book does relatively little to celebrate his contributions to research on nerve, many of his past and present collaborators in muscle research were able to attend the conference and all of those contributed to this volume.

Robert Stämpfli's contribution (Chapter 1) does, however, recall the days in Cambridge and Plymouth when he witnessed the great work of Hodgkin, Huxley, and Katz on the squid giant axon, and collaborated with Huxley in demonstrating saltatory conduction in myelinated nerve. This contribution arose by a curious circumstance. I asked Professor Stämpfli at short notice whether he would be prepared to make a speech at the dinner after the meeting. To my relief he accepted with alacrity, and to my surprise he told me he knew exactly what he would say. It turned out that he had written an article about Andrew Huxley in the early 1970s, though this had not been published. He gave the speech as promised, on an idyllic summer's evening in the garden of the Master's Lodge at Trinity.

In my own account of Andrew Huxley's research on muscle (Chapter 2), I soon abandoned the attempt to write a straightforward account of his work, as so much of it is covered in his own book *Reflections on Muscle,* and in his other review articles, and it became pointless to try to compete. I have tried to adopt a style that is midway between Robert Stämpfli's reminiscences and an account of his research, drawing extensively on conversations with his collaborators and with Andrew Huxley himself. I hope that this account will serve as a background to the various contributions in the body of the book and also entertain the reader.

Because of other commitments I was unable to work on this book much until the summer of 1990, and authors have brought their contributions up to date as of August 1990.

Stanford   R. M. Simmons
May 1991

1

A. F. Huxley: an essay on his personality and his work on nerve physiology

ROBERT STÄMPFLI

This article was originally written in 1970 for an Italian editor who intended to publish a book about the Nobel Prize winners for Physiology and Medicine (which has never been published). Having had the privilege of collaborating with Andrew Huxley for most of the period of his work with Alan Hodgkin which finally was awarded the Nobel Prize, a colleague thought that I probably had a very personal experience of this historical event and might be the right author to describe this period. In doing so, this chapter gives my personal impressions of the personality of Andrew Huxley during the early part of his and my own scientific career. It also provides an account of his contribution to research in nerve physiology which preceded his very successful work in muscle physiology and biophysics which is the main subject of this book.

Parentage and life before World War II

Andrew Fielding Huxley was born in Hampstead, London, on November 22, 1917. His father Leonard Huxley was a son of the famous nineteenth-century biologist, educator, and writer Thomas H. Huxley who, in his book *Man's Place in Nature,* first popularised the view that human beings evolved from other animals.

Andrew Huxley wrote in his curriculum vitae:

> My father Leonard Huxley was for a time a classics master at Charterhouse School and later took up a literary career, writing a number of biographies and being the editor of the *Cornhill*

magazine. The children of his first marriage included Sir Julian Huxley, the biologist, and Aldous Huxley, the writer. After his first wife's death, my father married Rosalind Bruce, and I am the younger of the two sons of this marriage. My father died in 1933.

I was educated at University College School (1925–30) and Westminster School (1930–5; King's Scholar); and went up to Trinity College, Cambridge, in 1935 with a major entrance scholarship. I had turned over to science from classics in 1932 and went to Cambridge expecting that my career would be in the physical sciences: I have always been mechanically minded, and I was inspired at Westminster by the physics teaching of the late J. P. Rudwick. I naturally took physics, chemistry and mathematics in my part I at Cambridge, but the rules required me to take another science, and I picked physiology, largely on the recommendation of an old friend B. Delisle-Burns, now of the Physiology Department, McGill University. I found Physiology interesting, partly for its subject matter and partly through contact with Adrian, Roughton, Rushton, Hodgkin and the late G. A. Millikan (all Fellows of Trinity) and others in the department, and decided to specialise in it. I spent 1937–8 doing Anatomy with the intention of qualifying in medicine, and 1938–39 doing the part II course in physiology. In August 1939, I joined Hodgkin at the Marine Biological Laboratory at Plymouth for my first introduction to research.

Until his undergraduate days, Huxley had been a rather shy boy and kept to himself, to his thoughts, and his books. He lived with his parents and his older brother in Hampstead, a protected family life. His two famous half-brothers Aldous and Julian were so much older that he considered them rather as uncles than as brothers. Up to the age of 15, when his father died, he had certainly learnt many things from him and had felt the importance of his grandfather's work, as his father had written a biography on Thomas H. Huxley – T.H. as he used to be called.

Alan Lloyd Hodgkin was at that time already established as a top-class experimentalist in electrophysiology. He had been one of Andrew Huxley's teachers at Trinity College. When Hodgkin took him to Plymouth for their first work in collaboration – Huxley's introduction to research – they actually made the discovery that formed the basis for the work that led to their Nobel award. Hodgkin had already performed several sets of experiments with K. S. Cole in the United States, using the "giant axon" of the squid described by the Oxford zoologist J. Z. Young. These animals possess fast-conducting, very thick nerve fibres which enable them to extrude seawater from their hollow muscular mantle system by a fast contraction and to withdraw suddenly by this squirting action from an attacking animal or any danger they recognise. These quick reactions, which have to occur at relatively low temperature, have led to the development of the thickest single nerve fibres known in the animal kingdom – reaching a diameter of about 1 mm. In Plymouth the species *Loligo forbesi* was readily available. Its giant fibres frequently reached a diameter of half a millimeter. They could be separated from the mantle muscle by careful dissection without losing their property of conducting fast nervous impulses if they were kept in seawater.

The experiment attempted by Hodgkin and Huxley consisted of measuring the viscosity of the axoplasm by dropping mercury down the inside of the vertically mounted fibre from the cut end. Having suspended the fibre vertically from the cannula, they realised that it would be possible to put a concentric capillary through the cannula and use it as an electrode, recording for the first time the absolute value of the membrane potential difference between the inside and the external seawater. They found in the resting fibre a negative value of about −50–60 mV, of the inside with respect to the outside. During the nervous impulse, elicited by electrical stimulation of the fibre, this membrane potential reversed, for a fraction of a millisecond, to a value of +45–50 mV (inside positive), and then recovered within about one millisecond to its original resting value. The "action potential" – that is, the electrical signal revealing excitation of the nerve membrane – consisted thus of an impulse of about 110 mV amplitude, starting from the negative resting potential towards zero and "overshooting" this value by 40–50 mV. Overshoots of action potentials over the resting potential value had been observed previously, but had never been accepted as an experimental fact, as all observations had been obtained with external electrodes. The observation of Hodgkin and Huxley (1939), obtained by measurement of the membrane potential with an internal electrode, was direct evidence which invalidated the theory of Bernstein (1902) which suggested that nervous activity consisted of a breakdown of the resting potential to zero. Bernstein had anticipated the origin of the resting potential. He assumed a selective permeability of the resting membrane for potassium ions, and the existence of a diffusion potential due to the differential concentration of potassium – about 20-fold higher in the interior as compared to seawater. A sudden loss of selectivity and an increase of permeability to all ions during the nervous impulse would temporarily bring the membrane potential to zero. For 30 years this was a widely accepted explanation of the action potential, since no one had succeeded in measuring its absolute value. The result of the experiment of Hodgkin and Huxley, which was confirmed one or two years later by the independent measurements of H. J. Curtis and K. S. Cole (1940, 1942) in the United States, meant a revolution in the physiology of excitation. However, this revolution was seriously delayed by the political events of the time. All investigations were hindered by the outbreak of World War II. During the first year of the war, Huxley was a clinical student in London. "But when medical teaching in London was stopped by air attacks", he says in his curriculum vitae, "I changed to work of more immediate application, and spent the rest of the war on operational research in gunnery, first for Anti-Aircraft Command and later for the Admiralty". Nevertheless, he was elected to a research fellowship at Trinity College, Cambridge, in 1941, which he took up at the beginning of 1946.

During World War II

During the war, his military duty seemed not particularly demanding for his well-trained brain. He decided to keep it alert by designing three-dimensional models of stellated forms of the regular polyhedra (during some of his free hours). They had, if possible, to be designed to be cut out from one single piece of cardboard. It was then folded and stuck together to give a three-dimensional model. He then sketched a different drawing of the model for each eye, so that each drawing could be looked at with the appropriate eye at the appropriate distance to give a stereoscopic image. All these models are masterpieces of invention and mathematical planning. They represent an example of the kind of exercise which he imposed on his brain in the moments of leisure.

This occupation reveals some of the particularities of Huxley's personality. He is by no means a man who does not appreciate relaxing – "popping off for a holiday" – as he calls it. But even in his free moments his remarkable brain needs occupation: whereas other people try to do nothing, he switches to some "good old mathematical problem" that requires logical treatment and a high degree of organised thought. I suppose that Huxley, during the war period, discovered his wonderful gift of a perfect intelligence and a perfect hand.

At Trinity College, Cambridge

His research fellowship at Trinity was given under war conditions, which meant that he did not have to submit a voluminous experimental and theoretical project to be admitted. During the war, Hodgkin and Huxley met several times to write up the full paper concerning their discovery of the overshoot of the action potential. Hodgkin, who wrote the paper almost entirely, had been wondering whether their observation could mean that during activity the membrane became permeable to sodium ions; but since researchers assumed that the higher resting permeability to potassium than to sodium was due to the smaller size of the hydrated potassium ion, it did not seem plausible to suggest that there was a "pore" that would admit sodium ions more freely than potassium ions. Furthermore, the intracellular sodium concentrations reported by other authors mentioned a ratio of only about 2:1 for the external/internal sodium concentration. In addition, Cole and Curtis had found an overshoot of about a 100 mV in one case, which made a diffusion potential for sodium unlikely. This explains why the full paper submitted to the *Journal of Physiology* on February 20, 1945 (Hodgkin & Huxley, 1945) does not even discuss such a possibility. (*Editor's note:* The background to this work is described more fully by Hodgkin, 1977.)

But the discussion of the overshoot continued between Hodgkin and Huxley, and also Bernard Katz, from the Biophysics Department of University College, who had come back from Australia after the war. Hodgkin had considered the possibility of dipoles in the membrane, which by turning around could possibly produce an overshoot. This hypothesis was shot down by Huxley because it could not be tested. They had started measuring the amount of potassium accumulating in a thin layer of fluid outside a crustacean nerve fibre. Hodgkin had also conducted a preliminary experiment (unpublished) with sodium-deficient solutions which, when applied to crustacean axons, decreased the amplitude of the action potential and seemed to argue in favour of a sodium mechanism for generation of the overshoot. He decided to do additional experiments with sodium-deficient solutions, in collaboration with Bernard Katz in Plymouth in 1947.

Marriage to Richenda Pease

Andrew Huxley had made the acquaintance of Richenda Pease, daughter of the geneticist Michael Pease, who worked for the Agricultural

Research Council in Cambridge. She had been to a dance given by the Barlow family after the war, and Andrew had been invited to it by his contemporary at Trinity, Andrew Barlow, a cousin of Richenda. Richenda's mother was a daughter of Lord Wedgwood, the famous Labour politician. Richenda had become an undergraduate at Newnham Women's College, where she studied physics, chemistry, and physiology. Huxley taught her in part I classes and was invited to Girton, where the Pease family had built a house.

Michael Pease, Richenda's father, had developed a reputation in the genetics of poultry by developing an autosexing breed, a particular cross where the plumage differs according to sex. Like his wife, he came from a family with considerable sympathy for Labour politics. His father Edward Pease was one of the founders of the Fabian Society.

Both parents liked to invite to their home a large number of young people. They had expanded their house, which had been built soon after the first war, by adding two first-war army-surplus huts to the existing house on their grounds. One of them contained a library and working room for Michael Pease. The other served as a children's room, dining room, and dancing floor according to the needs of the moment. They had a big garden, and in one of the trees the children had built a tree house, where Andrew and Richenda hoped to retreat from the family crowds. Richenda's brother, however, discovered them and tried to take away the steps that gave access to the tree house. This event was noticed by the family and marked the beginning of more and more frequent meetings and rising sympathy between Richenda and Andrew. They became engaged in the summer of 1946 and were married in July 1947. During their engagement, Andrew lived at Trinity College.

A presumed diffusion pathway for sodium

Hodgkin and Huxley frequently discussed the current–voltage relation of an excitable membrane and the physical basis of the permeability changes that were obviously involved in the generation of the action potential. Hodgkin who according to Huxley was far ahead in his understanding of physical chemistry (which is denied by Hodgkin), was actively elaborating theories of ways in which the permeability of a lipid membrane might vary with membrane potential. Huxley began to calculate action potentials under the assumption that depolarisation of the membrane produced an immediate increase of sodium permeability; but his first try showed that on this assumption alone, there would be a potential rise to a peak without recovery. Hodgkin proposed the presence of a negatively charged sodium carrier that might bind other cations, such as hydrogen, at the inside of the membrane after carrying sodium into the fibre. This would represent an inactivation process which could be included in the mathematical treatment.

In early 1947, Huxley began to compute action potentials, first without and then with inactivation of the presumed diffusion pathway for sodium ions which was opened by depolarisation of the membrane by an electrical stimulus to or above the threshold level. He computed membrane action potentials and also conducted action potentials. Each calculation took at least a week to ten days. He accomplished them mostly during his free hours. His computations made the work of both Hodgkin and Huxley tremendously exciting. Hodgkin wrote to me in a personal letter, "All the action potentials were computed by Andrew and I don't know anyone else who could have done them".

During this period, Hodgkin and Huxley also measured the potassium release due to activity of the nerve membrane by recording the membrane resistance (Hodgkin & Huxley, 1946). "In this I was very much a learner", admits Huxley. Hodgkin went to Plymouth for one month (June 20 to July 19), to do sodium experiments on squid fibres from July 4 to 12. These experiments gave clear evidence of the linear dependence of the overshoot on the logarithm of the external sodium concentration $[Na]_o$ and supported the idea, which had so far been rejected, of a specific pathway for sodium across the membrane – a pathway made available only during the action potential. Hodgkin included some of these results in his presentation of the Hodgkin and Huxley communication at the International Congress of Physiology, Oxford, entitled "Potassium leakage and absorption by an active nerve fibre" (Hodgkin & Huxley, 1947a,b). Huxley, who during this period had been on his honeymoon with Richenda, came to the Congress directly from Scotland.

Experiments with single myelinated nerve fibres at Cambridge

It is then that I met both Hodgkin and Huxley for the first time. I presented a paper on single nodes of Ranvier, and Professor Alexander von Muralt, my teacher in Berne, introduced me to them. Together with Professor E. D. Adrian, he had already planned a research stay of six months for me in Cambridge as a temporary guest in the Rockefeller

Unit at the Physiological Laboratory. Hodgkin had written a letter in April of the same year, suggesting that we use this stay to demonstrate saltatory conduction in myelinated fibres by showing that longitudinal currents within one internode occurred simultaneously, but showed a delay from one internode to the next: "This means that the time relations of an action potential recorded with a pair of electrodes spaced about 0.5 mm apart should not alter when the electrodes are slid along an internode. Hence instead of getting a straight line relation between conduction distance and conduction time one should get a series of steps". A drawing was added which corresponded very well to what Huxley and I found later during our common experimental work. The only prediction that was not yet adequate was the method suggested by Hodgkin, but, as will be shown later, he helped afterwards by proposing an important modification.

Hodgkin asked me to come first to Plymouth for a few days to see the work he was doing with Katz at that time. They were determining the temperature dependence of the squid action potential: "cooking fibres", as they called it.

After the Oxford Congress, Huxley was committed to two visits to introduce his wife to relatives – the reason why he was absent during some important first experiments to prove the sodium theory. As always in Plymouth during good weather, a trawler went out in the early morning to catch the fish and squid. Only in the early afternoon were the animals unloaded and put into the seawater tanks of the laboratories of the Marine Biological Association. The experiments began in the evening after careful dissection of the giant fibres of the biggest animals and went on for most of the night, often until the early morning hours of the next day. I was deeply impressed by the beauty of this experimental work and by the very critical attitude of Hodgkin and Katz. I was given a little side job – to try dissecting nerve fibres from the ventral cord of shrimps. I did my best but was not successful. Nevertheless, both Hodgkin and Katz were extremely kind and let me ask many questions, some of them showing clearly how little I knew and understood about the physiology of excitable membranes. During these few days in Plymouth, I began to realise how privileged I was to have an opportunity to work in England, where physiology was at its best and where these relatively young men represented the top researchers in this field. After approximately a week at the end of September, Hodgkin and I went to Cambridge, where I was expecting to meet Huxley at the Physiological Laboratory.

First impressions

When I went there in the early morning, after having become one of the guests at the Hermitage Guest House, and spent the first night in my room in an attic of Newnham Grange next door, the house of the physicist Sir Charles Darwin, I had great expectations of starting some fascinating work as soon as possible. However, at the department, nobody seemed to be "in", as I could gather from the board at the entrance, where all the people doing research were listed, from Professor E. D. Adrian to the newest postgraduate. I made myself known to Miss Elton, the secretary, and was told that most people would probably arrive at about 10 o'clock. Andrew Huxley arrived at this time in an excellent mood and took me to the basement of the building, where the rooms of the Rockefeller Unit occupied by Hodgkin, Huxley, and myself were. At the end of the corridor was E. D. Adrian's laboratory. I discovered later that it had a red light above the door, which was turned on when he was in. This meant that he was not to be disturbed. I was told that he escaped through the window if insistent visitors wanted to see him. At present, the light was not on; I was therefore allowed to have a glimpse of the many pieces of old-fashioned-looking apparatus, mostly rather untidy, and to see the dusty windows and the very big loudspeaker with a beautiful big trumpet.

Hodgkin, who had also arrived, tried an experiment with a single myelinated nerve fibre with me which was inconclusive. He was then very busy working out the results of his two consecutive stays in Plymouth. As Huxley got interested in dissecting single nerve fibres, Hodgkin then decided to put his laboratory at our disposal and to work mostly at home. We were thus able to use his excellent electronic equipment and Huxley's laboratory for dissection and for constructing nerve chambers. The most important tool for this was a foot-operated lathe which was Huxley's preferred instrument.

We began by building a suitable dissection stand for single nerve fibres with good dark-field illumination. Huxley, who turned out to be an expert optician, designed a dark-field condenser from a lens of an old war-surplus projector. Nylon threads that were used to tie single nerve fibres came from a parachute rope. When particularly fine threads of less than 20-μm diameter were required, we used the hair of Hodgkin's daughter Sarah, who was then a baby.

I soon realised the importance of the daily tea in the common room of the Physiological Laboratory. Huxley was usually hungry at this time of the day and stopped in the midst of an experiment when tea time came. We went to the common room, where a big kettle of water was boiling and tea was available. Several paintings on the wall, particularly the one of Sir Joseph Barcroft, gave a college atmosphere. Huxley, then about 30 years old, was known to come in regularly for tea and to eat "buns" with margarine and jam. All those in the laboratory and others from outside, who wanted to make use of his remarkable intelligence, came and waited respectfully until his second helping before asking questions. He would at first listen and continue chewing. Then, instead of answering directly, he usually reformulated the question much more precisely and to the point than others had been able to put it. He then gave a quick answer if the problem had become a pseudo-problem by his new formulation. But quite often he took a pencil and a sheet of paper and began to develop the adequate mathematical expression. The general belief of the audience was that no one could ever find a mistake in the work of his brain. This explained why so many who had difficulties getting their problems straight used Huxley as a human computer. Working with him was thus a great privilege. Not only I, but also Hodgkin and Katz, appreciated his unfailing logic and mathematical talent. On such occasions, Huxley not only proved to be a brilliant thinker, but also showed an amazing knowledge of biology, physics, and chemistry and an excellent memory as well.

Saltatory conduction

Our work went on very well. We got a good idea of how to measure longitudinal currents along a myelinated nerve fibre from Hodgkin. He came one morning with a piece of transparent plastic tape, folded in two. If oil was put in between the folded tapes, he argued, a vertical "oil gap" could be used to separate two compartments of saline. A nerve fibre could then be pulled through fine openings across the oil. The potential difference between two troughs of physiological solution on both sides of the oil gap would be proportional to the longitudinal current. By moving the fibre, one could measure the longitudinal current produced at any place by the conduction of a nervous impulse and conclude whether transmission of impulses is saltatory – that is, occurring by jumps from one node of Ranvier to the next, through the purely passive cable properties of the internodes – or continuous. The idea was excellent, but the transparent plastic tape, however, was inadequate.

Huxley and I started by building an oil gap with two coverslips of glass which had been perforated by drilling fine holes into them with a diamond drill.

These coverslips were kept at a small distance by spacers and pressed together between two troughs. The small space between them was filled with oil after having introduced a single nerve fibre transversely across the gap and the holes of the coverslips. The troughs and most parts of the apparatus were made of Perspex®; the foot lathe was operated by Huxley himself, who proved to be an expert mechanic. This has been confirmed by trained professional machinists, who affirmed that they had never seen anybody who could operate a lathe better than Huxley.

Within one month we had the first results, which clearly confirmed the theory of saltatory conduction. We discovered an inconvenience in the use of the oil gap: when the nodes passed through the gap, they sometimes remained covered with a film of oil. Thus we had to build another chamber. Using a single glass capillary of 50-μm diameter instead of the oil gap, we were able to obtain complete results for our first paper: "Evidence for saltatory conduction in peripheral myelinated nerve fibres". We finished our experiments around Christmas of that year; one of our typical experiments published actually carries the date of December 25, 1947. We did it after eating Christmas turkey at different places – Huxley with his in-laws at Girton, I with Adrian and his family! We wanted to write up the paper as soon as possible, but I had to return to Switzerland after Christmas. This is why I convinced Andrew and his wife Richenda, who was expecting her first baby, to come with me to Switzerland, where they could stay with us and where Andrew and I could finish the paper and present a communication to the Swiss physiologists (Huxley and Stämpfli, 1948). There was a big Christmas party in Girton where I also met Mrs. Huxley, Andrew's mother, who worried about Richenda's traveling by air in her condition. We decided nevertheless to fly to Switzerland. This enabled us to finish the paper and to submit it to the *Journal of Physiology* in June 1948 after very long and careful preparation (Huxley & Stämpfli, 1949a). This is where I discovered with what a critical eye Hodgkin and Huxley wrote their papers. Every word was weighed and considered in all its meanings. Huxley was particularly insistent on using the word "evidence" in the title, because at that time there had been results published by Tasaki which favoured saltatory conduction without, however, giving the kind of direct proof which, according to Huxley, deserved the term "evidence" only.

The Huxleys enjoyed their stay in Switzerland, where food rationing had ended, and returned gaily to England after a few weeks. A further stay in summer 1948 gave me the opportunity to see how the common work of Hodgkin and Huxley had been progressing since the end of 1947. Hodgkin and Katz had been working in Plymouth, setting up a voltage-clamp arrangement for single squid nerve fibres. According to all three (Hodgkin, Huxley, & Katz), this potentiostatic method was the best way to measure changes in membrane permeability to sodium and potassium as a function of time. One had only to impose a sudden depolarisation on the excitable membrane and, using an electronic feedback system, maintain this new potential rigorously constant, despite the permeability changes that would occur at the new potential. The currents produced by the potential change across the membrane would then be carried by ions and have no capacitative component except at the "make" and at the "break" of the potential change. These currents could then be interpreted as changes of ionic permeability of the membrane as a function of potential and time. By replacing sodium ions with non-penetrating cations such as choline, the sodium currents could then be separated from other ionic currents.

Hodgkin and Katz built this system after discussions with Huxley, who was then working with me in Cambridge. We intended to go on with our work on myelinated frog nerve fibres, attempting to measure the absolute value of resting and action potentials, and thus make possible a study of the influence of the external potassium and sodium concentrations on the resting and action potential. Huxley had to join Hodgkin and Katz towards the end of August in Plymouth, so that we had only two and a half months to develop a new method. The results of our efforts were meagre: we tried to use very long capillaries and introduce a single nerve fibre into them – a nerve-racking business in the true sense of the word, which was very depressing. We finally came to the conclusion that the only possibility was to build a new nerve chamber with four compartments, one of which was an oil gap having a total width of less than one internode. If a nerve fibre was pulled across the gap and arranged to have one node on either side, it was possible to detect the longitudinal current due to depolarisation of one node (by cutting a node or by applying an isotonic KCl solution to it) across the oil gap and oppose a voltage source to the flow of current across an adjacent small gap of lower resistance. The voltage that would bring the longitudinal current to zero would then be equal to the absolute value of the resting potential of the intact node. We intended to realize this project during my next stay in 1949.

Huxley joined Hodgkin and Katz in Plymouth and contributed some final touches to make the voltage clamp of the squid fibre work satisfactorily. The subsequent experiments in September 1948 were a great success. In the beginning, they had some difficulties in interpreting the current curves that were obtained, as none of the three had ever before used potentiostatic methods. But after a few days, they showed clearly that a sudden depolarisation of the nerve membrane induced an initial increase of the sodium permeability, demonstrated by an inward current of sodium or an outflow, if the depolarisation was larger than the thermodynamic equilibrium potential for sodium ions, calculated according to Nernst. Since at that time no accurate values of the internal sodium concentration in squid nerve existed, they could only conclude that the ratio of sodium concentration between the outside and inside was not 2:1, but about 10 or 20:1. This conclusion adequately explained the overshoot of the action potential. This observation was later confirmed by the activation analysis of squid axoplasm by Hodgkin and Keynes.

At a symposium on "Ionic Factors in Electrophysiological Events" held in Paris from March 31 to April 9, 1949, Hodgkin, Huxley, and Katz presented for the first time the general framework of the ionic theory of excitation to a selected audience of electrophysiologists (Hodgkin, Huxley, & Katz, 1949). K. S. Cole, one of Hodgkin's teachers in electrophysiology, was among them. Everybody realised that an outstanding contribution to a better understanding of nerve activity had been made by a very elegant piece of experimental work based on solid theoretical grounds. The amazing fact was that a few months before the actual experiments were carried out and the the paper was presented, the experimental data already had been evaluated, and an action potential computed on the basis of 23 equations, given in the theoretical section, revealing the efficiency of the teamwork. The mathematical part was elaborated mostly by Huxley, who had computed a great number of action potentials even before these voltage-clamp experiments were started!

As our work on saltatory conduction was already in press in the *Journal of Physiology,* we had no difficulty in giving an account of this paper as well, together with a small histological addition, once more owing to Huxley's efficiency (Huxley & Stämpfli, 1949b). He foresaw that one of the arguments of other electrophysiologists against saltatory conduction might be that, according to the textbooks of that time, myelinated nerve fibres of the central nervous system had no nodes of Ranvier. Having been a demonstrator in the histology course, Huxley decided to make methylene blue and osmium tetroxide stains of rabbit's spinal cord. With both methods, sections showed the existence of nodes, with similar internodal distances as in peripheral nerve. This report agreed with earlier findings of many famous histologists which had been forgotten for some time, owing to a few bad experiments of well-known histologists who did not succeed in staining the nodes. The good histological work of Huxley invalidated the argument based on such errors. In 1948, he had given to the Physiological Society a demonstration of his discovery, which was independently confirmed by three other papers in the same year.

From June 1949 until mid-August, Hodgkin and Huxley, and for a part of this time Katz, had been in Plymouth for another series of voltage-clamp experiments which showed clearly that the activated membrane pathway for sodium ions is inactivated during maintained depolarisation. The results of these experiments, which dealt mainly with current–voltage curves of the membrane of the squid axon, were not published until 1952. They then appeared with the four essential paper which were the basis of the Nobel award. For Huxley, 1949 was a prolific year. He was in Plymouth for two and one-half months, then "popped off" for a holiday in Scotland, and was back in Cambridge on September 30 to continue our work on absolute membrane potential measurements of myelinated nerve fibres. This time our method worked beautifully: we measured resting potentials and how they were influenced by changes in the external potassium concentration. Then we confirmed the sodium theory for myelinated nerve by showing that the overshoot is a linear function of $\log[Na]_o$.

Again we travelled to Switzerland to write up our new papers, this time via Paris, where we celebrated New Year's Eve. We enjoyed French food and wine as a remarkable contrast to the postwar austerity experienced in Great Britain. After arriving in Berne, we first prepared a short note to the Swiss Physiological Society (Huxley and Stämpfli, 1950b) and another for the International Congress in Copenhagen. (Huxley & Stämpfli, 1950a). We submitted the full papers to the *Journal of Physiology* in June. They were published in 1951 (Huxley & Stämpfli, 1951a,b). Hodgkin and Huxley also presented a paper on ionic exchange and electrical activity in nerve and muscle in Copenhagen (Hodgkin & Huxley, 1951). They continued their analysis of the experimental work in Plymouth from September 1949 to September 1951 and found full experimental evidence for the sodium hypothesis, by showing a potential- and time-dependent rapid increase of the sodium permeability followed by a slower decrease (inactivation). An increase in potassium permeability occurred only with a delay and a time constant similar to that of inactivation (Hodgkin & Huxley, 1952a,b,c; Hodgkin, Huxley, & Katz, 1952). The final and most difficult paper presented the mathematical model of membrane behaviour: "A quantitative description of membrane current and its application to conduction and excitation in nerve" (Hodgkin & Huxley, 1952d). Huxley contributed very much to this concluding paper: among other mathematical tasks, he had to do the numerical computation of action potentials. An electronic computer would have been available in Cambridge for this work, but only with a delay of six months. Huxley therefore decided to do the computations by hand. He thus calculated more than a million values on a hand-operated calculating machine, finishing in less than six months!

These fundamental five papers, together with the reviews given at symposia (Hodgkin & Huxley, 1952e), the Croonian lecture of Hodgkin (1958), and his book *The Conduction of the Nervous Impulse* (Hodgkin, 1964) undoubtedly were essential for the Nobel Committee in deciding that their work and that of Sir John Eccles should be jointly awarded the Nobel Prize in 1963.

But we are still in 1952. Hodgkin and Huxley measured, as the last piece of joint experimental work, the movement of radioactive potassium across the nerve membrane. It was mainly the amount of potassium leaving the fibre during maintained depolarisation that accounted for the observed delayed outward current. This was direct evidence that delayed currents were carried by potassium ions (Hodgkin & Huxley, 1952f). Already before finishing this last paper, Huxley, who had now proved himself to be the equal of his former supervisor, had turned to experimental problems of his own in muscle physiology.

Publications by A. F. Huxley on nerve in chronological order

Hodgkin, A. L., & Huxley, A. F. (1939). Action potentials recorded from inside a nerve fibre. *Nature, London, 144,* 710.

Hodgkin, A. L., & Huxley, A. F. (1945). Resting and action potentials in single nerve fibres. *Journal of Physiology, London, 104,* 176–95.

Hodgkin, A. L., & Huxley, A. F. (1946). Potassium leakage from an active nerve fibre. *Nature, London, 158,* 376.

Hodgkin, A. L., & Huxley, A. F. (1947). Potassium leakage and absorption by an active nerve fibre. *Abstracts of Communications at the 17th International Congress of Physiological Sciences,* pp. 82–3.

Hodgkin, A. L., & Huxley, A. F. (1947). Potassium leakage from an active nerve fibre. *Journal of Physiology, London, 106,* 341–67.

Huxley, A. F., & Stämpfli, R. (1948). Beweis der saltatorischen Erregungsleitung im markhaltingen peripheren Nerven. *Helvetica Physiologica et Pharmacologica Acta, 6,* C22–4.

Huxley, A. F., & Stämpfli, R. (1949). Evidence for saltatory conduction in peripheral myelinated nerve fibres. *Journal of Physiology, London, 108,* 315–39.

Huxley, A. F., & Stämpfli, R. (1949). Saltatory transmission of the nervous impulse. *Archives des Sciences Physiologiques, 3,* 435–48.

Hodgkin, A. L., Huxley, A. F., & Katz, B. (1949). Ionic currents underlying activity in the giant axon of the squid. *Archives des Sciences Physiologiques, 3,* 129–50.

Hodgkin, A. L., & Huxley, A. F. (1950). Ionic exchange and electrical activity in nerve and muscle. *Abstracts of Communications at the 18th International Congress of Physiological Sciences,* 36–8.

Huxley, A. F., & Stämpfli R. (1950). Direct determination of the membrane potential of a myelinated nerve fibre at rest and in activity. *Abstracts of Communications at the 18th International Congress of Physiological Sciences,* 273–4.

Huxley, A. F., & Stämpfli, R. (1950). Direkte Bestimmung des Membranpotentials der markhaltigen Nervenfaser in Ruhe und Erregung. *Helvetica Physiologica et Pharmacologica Acta, 8,* 107–9.

Huxley, A. F., & Stämpfli, R. (1951). Direct determination of membrane resting potential and action potential in single myelinated nerve fibres. *Journal of Physiology, London, 112,* 476–95.

Huxley, A. F., & Stämpfli, R. (1951). Effect of potassium and sodium on resting and action potentials of single myelinated nerve fibres. *Journal of Physiology, London, 112,* 496–508.

Hodgkin, A. L., & Huxley, A. F. (1951). Transport of radioactive potassium by current through a nerve membrane. *Journal of Physiology, London, 115,* 6P (title only).

Hodgkin, A. L., Huxley, A. F., & Katz, B. (1952). Measurement of current-voltage relations in the membrane of the giant axon of *Loligo. Journal of Physiology, London, 116,* 424–48.

Hodgkin, A. L., & Huxley, A. F. (1952). Currents carried by sodium and potassium ions through the membrane of the giant axon *Loligo. Journal of Physiology, London, 116,* 449–72.

Hodgkin, A. L., & Huxley, A. F. (1952). The components of membrane conductance in the giant axon of *Loligo. Journal of Physiology, London, 116,* 473–96.

Hodgkin, A. L., & Huxley, A. F. (1952). The dual effect of membrane potential on sodium conductance in the giant axon of *Loligo. Journal of Physiology, London, 116,* 497–506.

Hodgkin, A. L., & Huxley, A. F. (1952). A quantitative description of membrane current and its application to conduction and excitation in nerve. *Journal of Physiology, London, 117,* 500–44.

Hodgkin, A. L., & Huxley, A. F. (1952). Propagation of electrical signals along giant nerve fibres. *Proceedings of the Royal Society [B] 140,* 177–83.

Hodgkin, A. L., & Huxley, A. F. (1952). Movement of sodium and potassium ions during nervous activity. *Cold Spring Harbor Symposia on Quantitative Biology, 17,* 43–50.

Hodgkin, A. L., & Huxley, A. F. (1953). Movement of radioactive potassium and membrane current in a giant axon. *Journal of Physiology, London, 121,* 403–14.

Huxley, A. F. (1954). Electrical processes in nerve conduction. In *Ion Transport Across Membranes,* ed. H. T. Clarke. New York: Academic Press.

Huxley, A. F. (1959). Can a nerve propagate a subthreshold disturbance? *Journal of Physiology, London, 148,* 80–1P.

Huxley, A. F. (1959). Ion movements during nerve activity. *Annals of the New York Academy of Sciences, 81,* 446–52.

Huxley, A. F., & Pascoe, J. E. (1963). Reciprocal time-interval display unit. *Journal of Physiology, London, 167,* 40–2P.

Huxley, A. F. (1964). The quantitative analysis of excitation and conduction in nerve. *Nobel Lecture.* Stockholm: Nobel Foundation.

Huxley, A. F. (1964). The quantitative analysis of excitation and conduction in nerve (reprint of Nobel Lecture). *Science, 145,* 1154–9.

Huxley, A. F. (1964). Die quantitative Analyse der Nervenerregung und Nerven-leitung (translation of Nobel Lecture). *Angewandte Chemie, 15,* 668–73.

Frankenhäuser, B., & Huxley, A. F. (1964). The action potential in the my-elinated nerve fibre of *Xenopus laevis* as computed on the basis of voltage clamp data. *Journal of Physiology, London, 171,* 302–15.

Huxley, A. F. (1964). Nerve. In *Encyclopaedic Dictionary of Physics,* p. H78. Oxford: Pergamon Press.

Huxley, A. F. (1968). Effect of a local change of membrane resistance on the electrotonic potential recorded nearer to the source. Appendix to: Diamond, J. The activation and distribution of GABA and L-glutamate receptors on goldfish Mauthner neurones: an analysis of dendritic remote inhibition. *Journal of Physiology, London, 194,* 669–728.

Huxley, A. F. (1972). Research on nerve and muscle. In *Perspectives in Membrane Biophysics,* pp. 311–17. Chicago: Gordon & Breach.

Hodgkin, A. L., & Huxley, A. F. (1981). Abstract and commentary. (Citation Classic on the 1952 paper. *Journal of Physiology, London, 117,* 500–44). *Current Contents, 24,* 19.

2

A. F. Huxley's research on muscle

ROBERT M. SIMMONS

Light microscopy: sliding filament theory

Andrew Huxley switched from nerve to muscle research in 1951 immediately after completing the papers with Hodgkin on the squid giant axon. He later wrote, explaining the change: "For one thing I had never worked on anything but nerve, and for another, there was at that time (and indeed for a good many years after) no obvious way of pushing the analysis of excitation to a deeper level" (A. F. Huxley, 1977). Huxley's interest in muscle had been kindled when he took over the lectures on muscle to the final-year undergraduate students at Cambridge from David Hill (son of A. V. Hill, and himself a fine muscle physiologist). David Hill passed on his lecture notes, which included a description of some of the nineteenth-century observations of muscle using light microscopy. A number of authors had accurately described the appearance of striated muscle and the way in which the striations change during contraction, though there was a certain amount of conflict in the subsequent literature, particularly about which of the bands – the A-band or the I-band – changes in length when a muscle fibre is stretched or contracts (A. F. Huxley, 1977).

In 1951, knowledge of the structures underlying the striations was sketchy. It was thought that the muscle proteins actin and myosin existed as a complex that ran from end to end of a muscle, and theories of contraction involved a folding of this complex (e.g., Astbury, 1947). Early electron microscopy of muscle using longitudinal sections seemed to confirm this view, apparently showing that the filaments in the A-band and the I-band were continuous, though it was soon to be shown that myosin was located in the A-band. Huxley decided he would reinvestigate the changes in the striations that occur during contraction, mainly because he thought that therein was a clue to the underlying mechanism in the contraction bands that appear during active shortening. One thing to be discovered was whether these changes were related to activation, shortening, or tension. Besides, this study would also allow him to indulge a childhood enthusiasm for light microscopy.

The outlet for this interest was the construction of an interference microscope which was to enable him to measure the striations in a muscle fibre quantitatively and avoid the artefacts that bedevil the conventional microscopy of thick specimens. A low-power version was ready in 1951 (A. F. Huxley, 1952) and a high power version in 1952 (A. F. Huxley, 1954). By this time Huxley had been joined by Rolf Niedergerke, who had learned nerve dissection from Robert Stämpfli and who was now to apply the same skills to dissecting out frog single muscle fibres. This preparation was chosen because, as Ramsay and Street (1940) had shown, the fibres were relatively easy to keep alive. The choice of a living preparation was an obvious one for a physiologist to make, but it turned out to have significant advantages over simplified preparations – for example, ease of stimulation and regularity of the sarcomeres when activated – and Huxley continued to use it for most of his subsequent research.

The work proceeded at a fast rate. By early 1953, it was clear that the A-band remained sensibly constant under most conditions of passive stretch and active contraction. Huxley recalls that

> After we had seen the constancy of A-band width with the interference microscope, Rolf Niedergerke told me that W. Krause had described the same phenomenon in his book "Die motorischen Endplatten" published in 1869. This led me into a study of much of the 19th century literature on the microscopy of muscle, which I found highly interesting and instructive; it also gave me an immense admiration for many of the authors, especially Bowman, von Brücke, Krause, and Engelmann, all of whom were famous for work in other fields as well as muscle.
> (A. F. Huxley, personal communication)

It had been shown in the nineteenth century that the A-band had a high birefringence and therefore consisted of rodlets parallel to the fibre axis (i.e., the thick filaments). The results from one particular (and it turned out atypical) preparation suggested a sliding filament mechanism. During local stimulation with a slowly rising voltage, designed to produce slow shortening below the rest length, the fibre showed a set of contraction bands (the C_m bands) at the M-line in the middle of the A-band. This suggested that there was a second set of rodlets (i.e., the thin filaments) in the I-band which collided or overlapped at the M-line during shortening. However, progress was slow for the prosaic reason that during this period much of Huxley's time was taken up with his duties as press editor of the *Journal of Physiology* and secretary of the Council of Trinity College, Cambridge. Meanwhile at MIT, Hugh Huxley (whose Ph.D. in Cambridge had been on the muscle x-ray diffraction pattern) was looking at transverse sections of muscle in the electron microscope, and he demonstrated the now familiar double array of filaments (H. E. Huxley, 1953). Collaborating with Jean Hanson, Hugh Huxley then went on to make observations similar to those of Andrew Huxley and Rolf Niedergerke, but using myofibrils under phase-contrast microscopy.

The two sets of observations came to be published in the same issue of *Nature* (H. E. Huxley & Hanson, 1954; A. F. Huxley & Niedergerke, 1954) in the following way. Andrew Huxley met Hugh Huxley (the two are not related) and Jean Hanson by chance at Woods Hole in the summer of 1953. They had independently arrived at the sliding filament theory by this time. When Hugh Huxley was preparing his results for *Nature*, he wrote to Andrew Huxley to ask if he had anything in press to which he could refer, and as a consequence the papers were submitted and appeared together. This is not the place for a debate about which set of authors deserves the greater credit, though it is quite clear that Hugh Huxley had provided the ultrastructural evidence, the observation (possible in myofibrils but not in fibres) that the H-zone changes length when the sarcomere length changes, and the first statement of the sliding filament model (H. E. Huxley, 1953). However, it is remarkable that Andrew Huxley had got to much the same point, without the advantage of knowing the correct ultrastructure of the filaments.

As well as providing the structural basis for contraction, the two *Nature* papers also made some suggestions about the mechanism of force generation. Andrew Huxley and Rolf Niedergerke pointed out that if the relative force was generated at a series of points in the region of filament overlap, the isometric force should be proportional to the amount of overlap, and this was in agreement with the measurements of Ramsey and Street (1940). Hugh Huxley and Jean Hanson suggested three possible mechanisms. The first was that actomyosin linkages were formed as a consequence of ATP hydrolysis, and filament sliding was a consequence of a tendency for the number of such linkages to be maximised. The other two suggestions involved alterations of part of the length of the thin or thick filaments, which at that stage could not be ruled out.

The two Huxleys and their colleagues dominated the development of sliding filament theory and its ramifications. Hugh Huxley provided the evidence from electron microscopy which formed the basis of the cross-bridge mechanism of contraction. He wrote of one particular model: "Each myosin–actin linkage can pull the actin filament along a distance of, say, 132 Å by the contraction of a *branch* of the myosin molecule, the branch would then unhook and extend again until it linked up with the next molecule 132 Å along the actin filament" (Hanson & H. E. Huxley, 1955). Working in parallel, in 1954–5 Andrew Huxley developed a seminal theory of contraction which was to set the scene for much of the quantitative work on the contractile mechanism from about 1960 until the present (A. F. Huxley, 1957a). Apart from the reasoning about force generation that went into the 1954 *Nature* papers, the theory also incorporated the suggestion of Needham (1950) that the energetics of muscle (Hill, 1938) implied a cyclical process. The original article is demanding, and several people (including Brian Jewell and myself; Simmons & Jewell, 1974) have tried to provide simple introductions to it. But there is no substitute for the rigour of the original. A successful attempt to master the theory undoubtedly changes one's entire perception about how muscle works. It starts with very simple premises: (1) Muscle works by the relative sliding of the thick and thin filaments. (2) The sliding is caused by cross-bridges ("sidepieces" in the original) which project from the thick filament and interact with specific sites on the thin filament. (3) The cross-bridges have a limited range of movement. (4) They act cyclically, splitting a molecule of ATP per cycle. What then are the simplest assumptions about the kinetics that will explain the observed steady-state mechanical and energetic properties of muscle? The simplest assumption about force generation is that cross-bridges produce force by attaching only when they are extended. As long as the muscle is isometric, turnover must be low, so the detachment rate has to be made relatively slow. When a muscle shortens, the cross-bridges have to be made to detach more rapidly or the speed of shortening will be too slow. This leads to the concept of the dependence of the rates on cross-bridge position ("x-dependence" as it is now called, where x is the stretch or compression of the cross-bridge from its unstrained position). Although we know a lot more about the kinetics of the actomyosin ATPase than was known in 1953, the theory is still useful even in its original form.

Indeed its influence has been so all-pervasive that there is even a tendency for experiments and theorists to explain how their work relates to Huxley's 1957 theory, rather than to muscle itself.

Local stimulation: activation process

Huxley then turned his attention to the activation of muscle, again using light microscopy. A. V. Hill (1948, 1949) had shown that the rise of tension after a stimulus was too quick to be accounted for by the diffusion of a substance from the external membrane to the middle of a fibre. Therefore, there had to be some other mechanism to promote a more rapid inward spread of the depolarisation of the surface membrane. Hill pointed out that a process like the spread of crystallisation in a supercooled liquid "seeded" at a point could account for the results. As for a possible structure, the existence of internal space continuous with the exterior, thus implying an internal membrane system, had been demonstrated with the Golgi method by Veratti (1902). However, his discovery had gone largely unnoticed in an obscure journal (an English translation appeared in 1961). Early electron microscopy suffered from the lack of a reliable fixative for membranes, but by 1953 reticular structure between myofibrils had been observed and subsequently ascribed a role in the inward spread of activation (Bennett & Porter, 1953; Bennett, 1955).

Huxley was joined in Cambridge by Bob Taylor, the first of a long line of visitors from the United States. Their first set of experiments (unpublished) were designed to measure local shortening in response to depolarisation of fibres made inexcitable, using a "cross-roads" trough. They recorded shortening by means of light diffraction, with a mercury arc and recording onto a strip of film on a rotating drum. The aim was to measure the strength–duration curve, time course of activation, and so on. But they soon switched to a different approach, to test whether there was a local effect of activation at the Z-line. They used a micropipette to depolarise about 1 μm^2 of surface membrane (a similar experiment had been attempted by Gelfan in 1933, but he used too large a current, which led to irreversible contractures). They showed that a contraction occurred only when the tip of the micropipette was opposite the Z-line of a frog muscle fibre, moreover, that both adjacent half-sarcomeres always contracted together, whereas in crab muscle they could be excited separately. They went on to show that the stimulus was effective only at points 5 μm apart on the circumference of the Z-line. Huxley and Taylor concluded that "the influence of membrane depolarisation is conveyed to the interior of the fibre by spread along some structure in the I-band; from the anatomical situation this must almost certainly be Krause's membrane" (i.e., the Z-line in the sense of the entity extending across a fibre; A. F. Huxley & Taylor, 1955b). This left open the question of the actual mechanism, though Huxley subsequently discounted crystallisation or activated diffusion, as these would have entailed a propagated response and not a local contraction as observed (A. F. Huxley, 1957a). Besides, the structural evidence for a membrane-mediated response was becoming compelling.

However, there was a sticky moment at the meeting of the Physiological Society when Huxley first presented the results (A. F. Huxley & Taylor, 1955a). Quite unexpectedly, J. D. Robertson showed a slide of an electron micrograph from lizard muscle, showing the internal membrane system to be located not at the Z-line, but at the end of the A-band (Robertson, 1956). In the end, both results turned out to be right. In frog twitch muscle fibres the membrane system consists of a network of tubules surrounding each myofibril at the level of the Z-line; each half sarcomere of lizard muscle can be excited separately as in crab muscle (A. F. Huxley & Straub, 1958). (Ralph Straub was a Swiss postdoctoral fellow who worked with Huxley for a year in Cambridge. He died in 1988.)

Confusion over differences in preparations and the location and function of the internal membrane system prevailed for several more years until the advent of glutaraldehyde as a membrane fixative. (A blow-by-blow account of the development of the electron microscopy and physiology is contained in the review by A. F. Huxley, 1971.) Huxley taught himself electron microscopy (EM) and indeed went on to design an ultramicrotome (A. F. Huxley, 1957b). Some of his micrographs are published (A. F. Huxley, 1959), including those that showed for the first time that the T-tubule (the transverse tubule) is a continuous structure and not (artefactually) a row of separate vesicles.

Lee Peachey came to work with Huxley in 1958 (and collaborated subsequently again in Cambridge, at Columbia University, and more recently in London) to help resolve some of the issues over the location and function of the internal membrane systems. He was in fact still a student with Keith Porter (one of the leading electron microscopists in the fast-developing field of membrane structure) and had looked at the structure of frog slow fibres in the EM as part of the work for his thesis. Huxley and Peachey set about repeating the local stimulation experiments on frog slow fibres (Peachey & A. F. Huxley, 1960; A. F. Huxley & Peachey, 1964) and also did some further experiments on crab muscle which Peachey showed to have two sets of tubules, one at the Z-line and the other at the A–I junction, though only the latter set were functional (Peachey, 1967). The slow-fibre experiments were not immediately successful, and as Huxley and Peachey often incidentally isolated twitch fibres, they used them to reinvestigate the claims (e.g., Carlsen, Knappeis, & Buchtal, 1961) that tension generation was possible in fibres stretched to a sarcomere length at which there should be no overlap between the filaments. Huxley and Peachey confirmed this finding, but showed that the tension derived from comparatively inextensible sarcomeres at the ends of the fibres. When a fibre was allowed to shorten isotonically, no shortening occurred in the middle of the fibre unless there was some filament overlap (A. F. Huxley & Peachey, 1961).

The published records of the local stimulation experiments lose the impact of seeing them in real time, so while on a visit to Lee Peachey, Huxley put together a number of cinefilms at the Center for Mass Communication at Columbia University, with an introduction and commentary by himself, which was made available commercially. The film still serves as a useful teaching aid, showing the striations and their changes during contraction with great clarity and beauty.

In 1960, Huxley left Cambridge and moved to London to take up the appointment of Jodrell Professor of Physiology and Head of the Physiology Department at University College, London. This was a great centre for muscle research. A. V. Hill, the great pioneer of muscle mechanics and energetics studies, was still active, and Doug Wilkie and Roger Woledge were pursuing similar goals using chemical breakdown and heat measurements, respectively. Brian Jewell was making mechanical measurements on a number of types of muscle. The nearby Biophysics Department housed Bernard Katz, Ricardo Miledi, Gertrude Falk, Paul Fatt, Rolf Niedergerke, and Sally Page. For a while, Hugh Huxley was also in the Biophysics Department, though he soon returned to Cambridge, and the two Huxleys overlapped at University College for only one year.

There was one anomalous result from the studies by Huxley and Niedergerke (1954, 1958) of the A-band and I-band during contraction. Subsequent experiments, using the same method of stimulation that had first produced the C_m bands and suggested the sliding-filament mechanism, failed to show these bands and the constancy of the A-band; instead, both the A-band and I-band shortened. This continued to worry Huxley, and, very typically, he decided to reinvestigate the matter more thoroughly when Albert Gordon, a postdoctoral fellow from the United States, came to work with him in London. The problem was soon solved: the local stimulation conditions produced C_m bands at the edge of the fibre, but further into the fibre (where most of the observations were made) the myofibrils were inactive and were being shortened passively. Consequently, they were thrown into waves and the sarcomeres were viewed obliquely, giving false lengths for the sarcomere length and band widths (A. F. Huxley & Gordon, 1962), later confirmed by electron microscopy (Brown, Gonzalez-Serratos, & Huxley, 1984a,b).

Huxley continued his work on the activation of muscle, though principally through the work of visitors to his laboratory. It was by this time clear that vertebrate striated muscle fibres all have some kind of network of transverse tubules (the T-tubules or T-system) through which the depolarisation of the surface membrane is spread to the interior. The T-tubules make contact with the second internal membrane system, the sarcoplasmic reticulum (SR), in most muscles at structures known as "triads" from their appearance in the EM. Somehow the depolarisation of the T-tubule leads to the release of calcium stored in the SR, and the calcium activates the muscle filaments to produce force. After stimulation ceases, the SR reaccumulates calcium, causing relaxation. Much remained, however, to be discovered both structurally and otherwise about the membrane systems. In 1960, Walter Freygang had tried unsuccessfully in Huxley's laboratory to record the spike of current associated with the entry of the action potential into the T-tubule, using the same microcapillary electrode to stimulate a local contraction and to record the current, but the seal of the electrode was too leaky. It had also to be established that there really was continuity between the external solution and the lumen of the T-tubules. Visualisation of the opening of the T-tubule at the surface membrane in frog muscle proved to be difficult because the approach of a T-tubule to the surface is tortuous and is not usually visible in a single EM section. The situation is simpler in fish muscle, and Franzini-Armstrong and Porter (1964) were able to obtain convincing micrographs of the openings, using glutaraldehyde for the first time in muscle.

In frog muscle another approach was to show that some substance could penetrate inside the fibre, and Makoto Endo, a visitor to Huxley's laboratory from Japan, succeeded in doing this using a fluorescent dye (Endo, 1964). For the second time, work from the laboratories of the two Huxleys appeared in the same issue of *Nature,* as Hugh Huxley had demonstrated the continuity between the T-tubules and the external solution, using ferritin as a marker for electron microscopy (H. E. Huxley, 1964). Independently, Sally Page made the same observation as Hugh Huxley (Page, 1964), and David Hill also came to similar conclusions by measuring the fibre space available to albumin (Hill, 1964).

With the understanding of the geometry of the muscle membrane, the logical next step might have been to undertake a detailed study of the ionic processes, but Huxley did not do so, in part because a considerable effort was already being mounted primarily by Alan Hodgkin and his collaborators in Cambridge, but also because by now he had developed a strong interest in the mechanism of contraction. The study of the passive and ionic properties of the frog muscle membrane became an extremely intensive and challenging area, with contributions from Richard Adrian, Knox Chandler, Roy Costantin, Bob Eisenberg, Alan Hodgkin, Lee Peachey, and many others. This research led eventually to the discovery of a charge movement possibly involved in the mechanism linking the depolarisation of the T-tubule membrane to the release of calcium from the sarcoplasmic reticulum (Schneider & Chandler, 1973). For many years the molecular basis of this mechanism proved to be elusive, and the story of how it seems finally to have been solved is told by Clara Franzini-Armstrong in Chapter 4 of this volume.

Important studies complementing the research on the ionic mechanism of the surface and T-tubular membranes were measurements, in Huxley's laboratory, of the rate at which inward spread occurs. These experiments were done by Hugo Gonzalez-Serratos, a Ph.D. student from Mexico, using the simple but effective technique (derived from Huxley & Gordon's observations) of compressing a fibre in gelatin so that the myofibrils were thrown into waves, and filming them straightening as activation spread into the centre of the fibre (Gonzalez-Serratos, 1965, 1971). The rate of spread was fast, but it was still an open question as to whether the depolarisation spread passively down the T-system or was propagated in an action potential. The matter was finally settled by Constantin (1970), who showed that Huxley had for once drawn the wrong conclusion. In the local stimulation experiments, no action potential propagated across the diameter of the fibre through the T-system, but when a sufficient proportion of the T-system was depolarised (as would happen in a normal action potential) there was indeed a weakly propagating, sodium-dependent action potential in the T-tubules. An EM study was made by Lucy Brown (a staff member at University College) on the preparations used by Hugo Gonzalez-Serratos, eventually showing that the wavy fibrils developed at a range of sarcomere lengths when a fibre was compressed (Brown, Gonzalez-Serratos, & Huxley, 1984a,b) and not, as had first been thought, simply when the tips of the filaments met (Brown, Gonzalez-Serratos, & Huxley, 1970). The later work also includes a quantitative account of the bending of the filaments and the restoring forces. The approach of observing wavy myofibrils was later adopted by Reinhardt Rüdel and Stuart Taylor, postdoctoral fellows who worked together in Huxley's laboratory on the inactivation in the centre of a fibre which occurs when shortening proceeds below the slack length, presumably because the T-tubules became distorted (Taylor & Rüdel, 1970; Rüdel & Taylor, 1971).

After the depolarisation of the T-tubules, the next stage in the activation of muscle is the release of calcium and its reuptake during relaxation. Saul Winegrad spent a sabbatical year at University College, thinking out, with advice from Huxley, how to make measurements of the movement of calcium from one internal location to another. On his return to the United States, he proceeded to put theory into practice, using the autoradiographic technique he had developed in London (Winegrad, 1965). This marked the end of Huxley's involvement in research into the activation of muscle, though several of his collaborators went on to make some of the fundamental discoveries in these areas: for example, Makoto Endo on the release of calcium in skinned fibres (Endo, 1977; Chapter 5); and Reinhardt Rüdel and Stuart Taylor on the time course of myoplasmic calcium concentration after an action potential, using the luminescent protein aequorin (Taylor, Rüdel, & Blinks, 1975; Blinks, Rüdel, & Taylor, 1978).

Muscle mechanics: cross-bridge theory

Huxley next turned to a study of the mechanics of muscle fibres, a study that was to continue to occupy him until the present time. The work of Ramsey and Street (1940) on the length–tension relation in single fibres had shown that there was a qualitative dependence of tension on filament overlap as predicted by the sliding filament theory, but not the quantitative relationship that was needed to prove the point. Huxley suspected that this discrepancy was due to the tension produced by inextensible sarcomeres at the ends of the fibres, which he and Peachey had previously demonstrated. His solution to this problem was to devise a method of recording the length of a segment at the middle of the fibre, and to use the signal in a feedback system to keep the segment length constant. The tension produced would then be appropriate to the isometric level of the centre of the fibre, with the regions at the ends being allowed to shorten automatically by a motor attached to one end of the fibre.

Measurement of segment length could have been achieved in two ways. One would have been to record the optical diffraction pattern and hence sarcomere length (a method that Huxley and Taylor had previously used). This has the advantage of being noninvasive but the disadvantage that, if the fibre moves along in the light beam, there is an artefactual change of recorded sarcomere length if the newly illuminated part has a spacing that is slightly different from that of the part moving out of the beam. Huxley chose instead to attach two markers to the fibre and to measure the distance between the markers by a "spot-follower system", in which the images of two spots on a cathode ray tube were focused one onto the edge of each marker. This worked well, at least for the speeds of movement to be used in the next 15 years, though the system was eventually to be supplanted by another method with a faster response time. Much of the spot-follower system and the associated servo-system and its electronics were built by Albert Gordon and Fred Julian (also a postdoctoral fellow from the United States), though in fact the two did not overlap in Huxley's laboratory and only met for the first time some five years after the famous papers that bear their name appeared (Gordon, Huxley, & Julian, 1966a,b). This work showed, first, that isometric tension was proportional to overlap between the filaments on the descending limb of the curve, and diminished in two phases on the ascending limb, at first shallowly (and probably due at least in part to decreased activation, as Rüdel and Taylor were to show). The start of the second, steeper part of the curve coincides with the length when the thick filament meets the Z-line. They also showed that the maximum velocity of shortening is independent of overlap. The picture to emerge from this work on the descending limb was simplicity itself. The cross-bridges act independently. The maximum velocity relates to the kinetics of the cross-bridge cycle, and not to the number of cross-bridges that are active. However, there were a number of loose ends, and some of these still have not been tidied up. The questions as to the exact lengths of the filaments and how these match up with the length–tension curve have never been resolved entirely satisfactorily (Page, 1968; Craig & Offer, 1976), and there is the vexed question of the "creep", the fact that after the initial rise of tension, there is a slower rise to a much higher level. This was shown by Gordon, Huxley, and Julian to be due in part to the sarcomeres at the ends, and it was assumed that the remainder was due

to the development of nonuniformities as the stronger sarcomeres shortened; but for many years this position was attacked by Gerald Pollack
on the grounds that he could apparently find fibres where no obvious
nonuniformity could be detected. However, he recently seems to have
capitulated, at least in part (Granzier & Pollack, 1990). Pollack has
attacked a number of other points of the current dogma, with an at times
spirited discussion in and out of the literature involving Huxley (A. F.
Huxley, 1984, 1986).

The next step in the mechanical studies was to look at the transient
properties. Huxley's cross-bridge theory of 1957 had dealt with shortening at a steady velocity, but there were clear implications for what would
happen if during an otherwise isometric contraction the length or force
was suddenly changed, because with finite rates of attachment and detachment the cross-bridges would take a finite time to adapt to the new
mechanical situation. Richard Podolsky and his colleagues at NIH were
the first to produce experimental evidence of this kind, applying stepwise changes of tension to a contracting muscle and observing "velocity
transients" preceding the steady-state velocity (Podolsky, 1960; Civan &
Podolsky, 1966). In order to account for their results, Podolsky and his
colleagues produced a modified version of the A. F. Huxley (1957a)
theory, with very different rate constants from the original (Podolsky,
Nolan, & Zaveler, 1969; Podolsky & Nolan, 1971). A feature of the
1970s was a friendly rivalry between the laboratories of Huxley and
Podolsky, which undoubtedly stimulated progress both experimentally
and theoretically.

Fred Julian and also Clay Armstrong (who spent two years in Huxley's
laboratory) did a large number of experiments that were not published
at the time, mainly because the equipment was not quite fast enough,
but they certainly deserve a large share of the credit for the subsequent
work on tension transients, and also for quite a lot of work that was
never fully published, because at the time it was hard to make a coherent story out of it. This work included the "pull-out" experiment, isotonic transients, and imposed oscillations to measure stiffness (some of
these experiments are referred to in reviews by A. F. Huxley, 1971,
1974, and by A. F. Huxley & Simmons, 1973). There was, however, one
brief publication from Fred Julian and Clay Armstrong about the tension transients and how these related to the isotonic transients, showing
that Huxley had now gone on to consider in detail the cross-bridge cycle
(Armstrong, Julian, & Huxley, 1966).

Clay Armstrong also did some mechanical experiments on crab fibres that have very long sarcomere lengths. In the 1954 *Nature* paper, there is a prediction about the mechanical properties of such fibres (A. F. Huxley & Niedergerke, 1954a) but in practice these experiments turned out to be problematic because, as Clara Franzini-Armstrong showed by EM in Huxley's laboratory, there is wide variability in the filament lengths, even in the same fibre (Franzini-Armstrong, 1966, 1970b).

During this period, the philosophy of applying length or tension steps was established, and so also was the idea of the "sarcomere length clamp" by analogy with the nerve voltage clamp. The idea was that if the length was suddenly changed by a small amount during a tetanus, the kinetics of the cross-bridge cycle would be revealed in the tension record, just as the kinetics of the ion channels had been revealed in the time course of the current after a sudden change of voltage. However, it was not entirely clear whether it was better to do length steps (resulting in tension transients) or tension steps (resulting in velocity transients). Length steps were in principle probably the right thing to use, but there was a lot to be said for tension steps, as any series compliance in the fibre would remain constant after the step.

My own contribution to this history started in 1968, when I joined Huxley to continue the mechanical experiments. I found the electronic equipment, which seemed to be mainly war surplus, pretty daunting to begin with. The first thing I switched on exploded in my face, but I gradually learned to master the equipment and even contributed some circuits of my own. I was recently amused to read Alan Hodgkin's comment that

> the objective in designing electronic equipment was not to make some neat miniaturised unit but to build up as massive and imposing an array of racks and panels as you could get – possible with the idea of cowing your scientific opponents or dissuading your rivals from following in your footsteps. (Hodgkin, 1977)

Huxley later referred to this equipment, which was housed in a single very tall rack, in terms of geological strata. There was the Precambrian of Al Gordon's spot follower, the Palaeozoic of Fred Julian's servo system, the Mesozoic of my tension transducer, and the Cenozoic of Lincoln Ford's motor. One day the Palaeozoic was left on a waste bin while we tested some new amplifiers in its place in the rack, and the cleaner threw it away. Next day a notice went up on the door of the laboratory: "NO CLEANING IN HERE".

We had to build a new tension transducer, as it was clear we were missing some of the earliest response when making rapid length or tension steps (A. F. Huxley & Simmons, 1968). The first experiments we did were to try to improve on the time resolution of the type of tension step experiment that Civan and Podolsky (1966) had published, and we also tried to couple these with a measurement of stiffness by imposing an oscillation. The stiffness certainly decreased markedly and rapidly after a tension step (as predicted by Huxley's 1957a theory), but we were worried that this might be due to some passive nonlinear compliance in the fibre. After a great deal of effort, we succeeded in producing reasonable steps by using a shape-follower which generated a waveform by an oscilloscope trace following the outline of a piece of cardboard. Huxley would cut out the cardboard with a pair of scissors to the required shape calculated by deconvoluting the observed response. But we finally turned to doing length steps.

I certainly had no clear picture about what we were looking for at this time. While I was aware of the various cross-bridge theories, much of the literature on muscle (and nonmuscle) mechanics was presented in terms of viscoelastic Voigt and Maxwell elements, and it was not at all clear how one could distinguish the passive properties of the material of a fibre from some interesting cross-bridge kinetics. Huxley clearly felt that it was preferable to experiment unfettered by prior interpretations, especially if these were likely to be incorrect. "You have to realise that the first law of muscle physiology is that X is always wrong", he once said in some exasperation at my continual worries about what was in the older literature. Huxley's approach seemed to be to try to do the best possible experiment and then worry about the interpretation later. I think he was completely convinced in an entirely intuitive way that there must be something there that would tell us how muscle works. We had a succession of working hypotheses and things to test, some of them based on "classical" mechanics and others on ideas about cross-bridge kinetics, but the major thing we should be looking for, I gathered, was some nonlinearity: for example, if a cross-bridge had a limited range of movement, there should be some sign of it in the mechanical response. This was one of the reasons why Huxley was hostile to the use of sinusoidal analysis. "Sinusoidal analysis only works in the linear range, and the interesting things about cross-bridges lie in the nonlinearities".

There were indeed hints of something interestingly nonlinear in the earliest part of the isotonic transients, but it was only when we turned to recording tension transients and plotted out the relation between length and tension at early times that it became obvious that the data could be interpreted in terms of an element with a restricted range of movement, which might correspond to the finite rotation of the cross-bridge head in Hugh Huxley's model of force generation (H. E. Huxley, 1969). Everything then fell into place, though our first conclusion was coloured by the prejudices we had developed that there must be a considerable amount of series compliance inside the sarcomere, and also that the cross-bridge was a pretty rigid object which rotated to produce force with little "play" in it (A. F. Huxley & Simmons, 1970). Experiments comparing the response at different degrees of overlap soon showed that this was wrong; in fact, the whole of the early response scaled with overlap and therefore came from the cross-bridges (A. F. Huxley & Simmons, 1971a). This picture seemed perfect, except for one important fact: the rate constant of early tension recovery showed a pronounced nonlinearity with respect to step size which was in the wrong direction for a simple viscoelastic element. Huxley solved this problem by showing that this sort of behaviour was to be anticipated for a transition from one defined state to another when the rate of the transition is affected by the force on the cross-bridge (A. F. Huxley & Simmons, 1971b).

There were mixed reactions to the theory. On the one hand was the view that the attachment–detachment cycle of A. F. Huxley (1957a), plus the structural picture of H. E. Huxley (1969), plus the biochemical cycle of Lymn and Taylor (1971), plus the force-generating step of A. F. Huxley and Simmons (1971b) equalled the muscle problem solved. We found this astonishing, as we knew we were just scratching the surface of the problem. On the other hand, there was quite a lot of hostility both in public and in print (e.g., Abbott, 1972; Blangé, Karemaker, & Kramer, 1972). I remember finding some of the more public debates about the theory, in what were my first important seminars, distinctly alarming, and I suspect that I collected a few brickbats intended for my illustrious colleague. Subsequently, Terrell Hill developed a general formalism for cross-bridge models (Hill, 1974, 1975), deriving the rules that a force-dependent (or "x-dependent") system must obey, and with Evan Eisenberg he went on to produce some examples based on the solution kinetics of the actomyosin ATPase, which Eisenberg, David Trentham, and Ed Taylor were working out at that time. It has to be said that Huxley did not like one of the examples, which obscured the simplicity of the physical basis for the rate constants in our original theory, and was wont to refer to it as "Huxley & Simmons emasculated".

Another alarming experience happened to me at about this time as a result of my work with Huxley: a meeting with A. V. Hill. The results that we had been getting on the rapid transient properties of muscle fibre carried some implications for the so-called active state of muscle, which A. V. Hill had invented to describe the rise and fall of activity of a muscle when it was stimulated. I took an extreme view of this, and with all the enthusiasm of a new convert preached against the very concept of the active state (Simmons & Jewell, 1974). Huxley wisely steered clear of an area where only fools rush in. By this time A. V. Hill had retired from University College, London, and was living in Cambridge. Before long, I found myself seated opposite him at a dinner in Cambridge and was told by Huxley, sitting next to me, to explain myself. I found this, to say the least, unsettling. For one thing, A.V. (as he was known) had a terrifying reputation in public debate, and many people in the field bore scars from the field of conflict with him. For another, while I genuinely believed the active state hypothesis was badly flawed, I had a profound admiration for the man who had put together the beautiful empirical theory that related mechanics and energetics (Hill, 1938). It was difficult to convey in conversation the gist of both these sentiments, but I did my best, explaining how the concept of the active state had been overtaken by the interpretation of muscle mechanics in molecular terms. He listened quietly, then at the end said, "Oh well, we didn't know much about molecules in those days". And then, with a piercing look at Huxley which I fancy combined admiration and a certain amount of envy, he said "Andrew is good at that sort of thing".

Although the theory and several of the experimental observations were by now published, we were far from happy about the quality of the data. Our apparatus was still much too slow, and we were fortunately joined at this time by Lincoln Ford, who proceeded to build a faster motor (Ford & Huxley, 1973), and we generally tuned up everything else as much as possible. Then we "settled down" (to use Huxley's phrase) to do the main experiments, looking at tension transients as a function of overlap of the filaments, during shortening and during the rise of tension. Huxley built a delay line for simulating the response and proceeded to a typically thorough analysis, but the work turned out to be extremely demanding and was regrettably published only after some considerable time, as we were all engaged on other things (Ford, Huxley, & Simmons, 1977, 1981, 1985, 1986).

This was by no means the only work on muscle in Huxley's laboratory during this time. Goodwin Breinin, a distinguished clinical and experimental ophthalmologist from New York, collaborated with Huxley in recording the contribution of slow, nonspiking fibres in the external eye muscles using both human subjects and anaesthetised cats. Nonspiking activity was detected during weak contractions and was not associated with vergence movements as had been supposed. Lydia Hill, a Ph.D. student, looked with the interference microscope at what happens to the sarcomeres when muscle fibres are stretched by several percent during a tetanus, to try to settle a debate about the origin of the residual tension that persists after the stretch (Hill, 1977). She found that the striations and the bands remain regular. However, she and Lucy Brown then went on to fix fibres rapidly both during the tension "creep" and after large stretches, and showed by EM that considerable displacements of the A-band from the centre of the sarcomere and also local gross irregularities occurred, which might account for the anomalies in tension (Brown & Hill, 1991; Chapter 13, this volume). Jan Lännergren, from the Karolinska Institute, spent a sabbatical year in the laboratory, and I recall being very impressed by his skills at dissection and making small pieces of equipment. He looked to see if the UV-absorbing bands in the sarcomere change during contraction (as would happen, for example, if part of the ATP were compartmentalised), but the results were negative (Lännergren, 1977).

Huxley became President of the Royal Society in 1980, and returned to Cambridge in 1984 to become Master of Trinity. This left him with little time for laboratory work, though he continued to develop new methods for muscle mechanics, collaborating with Vincenzo Lombardi and Lee Peachey, before he left London, in building a "striation follower" which ingeniously used groups of striations as natural markers, and gave a much better frequency response and less noise than the old "spot-follower" (A. F. Huxley, Lombardi, & Peachey, 1981). He has recently developed a new approach to understanding the diffraction pattern from muscle, regarding the sarcomeres as waveguides (A. F. Huxley, 1990). He has also continued to take an active role at the major conferences in the field – the Gordon Research Conference and the Alpbach Workshop – summarising the whole of the week's proceedings on the last day and indicating the way forward. His performance at scientific meetings is nothing short of legendary ("there is an inconsistency in the fourth panel on the left-hand side of your seventh slide"), and he even gave a lecture in Japan in Japanese, which he had been learning for fun. (Makoto Endo's account of the circumstances of this lecture appears in Chapter 5.)

Inventiveness and inventions

What are the main characteristics features of Andrew Huxley's research work that make it so distinctive? There are many testimonials, in this book and in the literature, to his phenomenal power in applying mathematical and physical techniques, but as he himself says, "It is not the mathematics that is difficult in biology, it's formulating the problem in the right way". He certainly has been careful to choose areas that are suitable for the strictly quantitative treatment. "It's a bog" is a favourite comment on complex preparations. He has also been highly selective about what he publishes, refusing to produce a full paper until he thoroughly understands the results, and even then refraining from publishing if a "good story" does not emerge. Proficiency at designing and building equipment is another characteristic. The famous Huxley micromanipulators (A. F. Huxley, 1961) are currently enjoying a new vogue among electrophysiologists and manufacturers; his ultramicrotome (A. F. Huxley, 1957b) was produced commercially by Cambridge Instruments; parts of the "striation follower" (A. F. Huxley, Lombardi, & Peachey, 1981) are being produced by Huxley's son Stewart (who also makes the micromanipulators); and there must be dozens of other devices (e.g., A. F. Huxley, Kearney, & Purvis, 1962; A. F. Huxley & Read, 1979), many in use in the laboratories of former collaborators throughout the world.

Huxley wrote a number of scholarly accounts of the history of the observations on the striations (in particular, A. F. Huxley, 1977, 1980). One of the lessons is that as the light microscope became a commonplace instrument, users forgot or never learned the correct way to use it. The importance of understanding measuring instruments and their limitations was drilled into his collaborators: one of the first things I was required to do was to check that the oscilloscope, which had not been used for some time, was working correctly, though I would have regarded it as a standard instrument. In fact, there was something wrong with it, and I had to put it right. This attitude made life uncomfortable for salespeople, especially when they saw their precious demonstration model being stripped down and the parts examined minutely. Huxley sent one microscope saleman from a famous firm packing with what he regarded as a substandard objective which a colleague had bought on Huxley's recommendation. The salesman was heard to complain as he left: "If we made all our objectives to the same standard as the ones we have to supply to Huxley, we'd be out of business."

As for influences, as Robert Stämpfli relates in Chapter 1, a physics master was a major influence in determining Huxley's interest in science, and clearly Alan Hodgkin was another. When, not for the first time, Huxley found my knowledge of mathematical methods somewhat wanting (I think it was the p-operator: "I can't see how you can manage without knowing that"), he added, "I learned that from Alan Hodgkin – like everything worthwhile I know about science". It is perhaps fitting to come full circle and quote from Hodgkin's own account of their work on the squid giant axon, which illustrates, above all, the extraordinary high standards they set themselves and which Huxley has continued to hold in his research on muscle contraction:

> We had started off to test a carrier hypothesis and believed even if that hypothesis was not correct we should be able to deduce a mechanism from the massive amount of data we had collected. These hopes faded as the analysis progressed. . . . [I]t became clear that the electrical data would by themselves yield only very general information about the class of system likely to be involved. (Hodgkin, 1977)

Most people would not consider their astonishing achievement to be a disappointment, but the epitome of the quantitative approach.

Publications on muscle by A. F. Huxley in chronological order

Huxley, A. F. (1952). Applications of an interference microscope. *Journal of Physiology, London, 117*, 52–3P.

Huxley, A. F. (1952). Measurements on diffraction spectra of single muscle fibres. *Journal of Physiology, London, 117*, 53P (title only).

Huxley, A. F. (1953). Observations with an interference microscope. *Biological Bulletin, Woods Hole, 105*, 390–1.

Huxley, A. F. (1954). A high-power interference microscope. *Journal of Physiology, London, 125*, 11–13P.

Huxley, A. F., & Niedergerke, R. (1954). Interference microscopy of living muscle fibres. *Nature, London, 173*, 971–3.

Huxley, A. F., & Niedergerke, R. (1954). Measurement of muscle striations in stretch and contraction. *Journal of Physiology, London, 124*, 46–7P.

Huxley, A. F., & Taylor, R. E. (1955). Activation of a single sarcomere. *Journal of Physiology, London, 130*, 49–50P.

Huxley, A. F., & Taylor, R. E. (1955). Function of Krause's membrane. *Nature, London, 176*, 1068.

Huxley, A. F. (1956). A new muscle preparation: isolated fibres from the crab. *Journal of Physiology, London, 135*, 35P (title only).

Huxley, A. F. (1956). Local activation of striated muscle from the frog and the crab. *Journal of Physiology, London, 135*, 17–18P.

38   Robert M. Simmons

Huxley, A. F. (1956). Present knowledge of the structure of striated muscle and its relation to function. *XX International Congress of Physiological Sciences Reviews*, pp. 219–23.

Huxley, A. F. (1956). Light microscopy of muscle. *British Medical Bulletin, 12*, 167–70.

Huxley, A. F. (1957). Muscle structure and theories of contraction. *Progress in Biophysics and Biophysical Chemistry, 7*, 255–318.

Huxley, A. F. (1957). An ultramicrotome. *Journal of Physiology, London, 137*, 73–4P.

Huxley, A. F. (1957). Das Interfernz-Mikroskop und seine Anwendung in der biologischen Forschung. In *Verhandlungen der Gesellschaft deutscher Naturforscher und Ärtze*, pp. 102–9. Berlin: Springer.

Huxley, A. F., & Niedergerke, R. (1958). Measurement of the striations of isolated muscle fibres with the interference microscope. *Journal of Physiology, London, 144*, 403–25.

Huxley, A. F., & Straub, R. W. (1958). Local activation and interfibrillar structures in striated muscle. *Journal of Physiology, London, 143*, 40–1P.

Huxley, A. F., & Taylor, R. E. (1958). Local activation of striated muscle fibres. *Journal of Physiology, London, 144*, 426–41.

Huxley, A. F. (1959). Local activation of muscle. *Annals of the New York Academy of Sciences, 81*, 446–52.

Huxley, A. F., & Peachey, L. D. (1959). The maximum length for contraction in striated muscle. *Journal of Physiology, London, 146*, 55–6P.

Huxley, A. F., & Peachey, L. D. (1960). Local activation of slow fibres from striated muscle of the frog. *Journal of Physiology, London, 151*, 43P (title only).

Peachey, L. D., & Huxley, A. F. (1960). Local activation and structure of slow striated muscle fibers of the frog. *Federation Proceedings, 19*, 257.

Huxley, A. F. (1960). Appendix to Solomon, A. K. Compartmental methods of kinetic analysis. In *Mineral Metabolism*, Vol. 1, Part A, ed. C. L. Comar & F. Bronner, pp. 119–67. New York: Academic Press.

Huxley, A. F. (1961). A micromanipulator. *Journal of Physiology, London, 157*, 5–7P.

Huxley, A. F., & Peachey, L. D. (1961). The maximum length for contraction in vertebrate striated muscle. *Journal of Physiology, London, 156*, 150–65.

Huxley, A. F. (1962). Skeletal muscle. In *Muscle as a Tissue*, ed. K. Rodahl & S. M. Horvath, pp. 3–19. New York: McGraw-Hill.

Huxley, A. F., Kearney, A., & Purvis, C. (1962). A dissecting microscope. *Journal of Physiology, London, 162*, 42–4P.

Huxley, A. F., & Gordon, A. M. (1962). Striation patterns in active and passive shortening of muscle. *Nature, London, 193*, 280–1.

Peachey, L. D., & Huxley, A. F. (1962). Structural identification of twitch and slow striated muscle fibers of the frog. *Journal of Cell Biology, 13*, 177–80.

Gordon, A. M., Huxley, A. F., & Julian, F. J. (1963). Apparatus for mechanical investigations on isolated muscle fibres. *Journal of Physiology, London, 167,* 42–4P.

Gordon, A. M., Huxley, A. F., & Julian, F. J. (1964). The length–tension diagram of single vertebrate striated muscle fibres. *Journal of Physiology, London, 171,* 28–30P.

Peachey, L. D., & Huxley, A. F. (1964). Transverse tubules in crab muscle. *Journal of Cell Biology, 23,* 70–1A.

Huxley, A. F., & Peachey, L. D. (1964). Local activation of crab muscle. *Journal of Cell Biology, 23,* 107A.

Huxley, A. F. (1964). Muscle. *Annual Reviews of Physiology, 26,* 131–52.

Huxley, A. F. (1964). Introductory remarks [Discussion meeting]. *Proceedings of the Royal Society [B], 160,* 434–7.

Huxley, A. F. (1964). The links between excitation and contraction. *Proceedings of the Royal Society [B], 160,* 486–8.

Huxley, A. F., & Julian, F. J. (1964). Speed of unloaded shortening in frog striated muscle fibres. *Journal of Physiology, London, 177,* 60–1P.

Elliott, G. F., Huxley, A. F., & Weis-Fogh, J. (1965). On the structure of resilin. *Journal of Molecular Biology, 13,* 791–5.

Huxley, A. F. (1965). Shinkei dendō no butsurigaku. *Kagaku, 35,* 601–7.

Huxley, A. F. (1965). Muscle tension on the sliding filament theory. *Excerpta Medica International Congress Series, 87,* 383–7.

Gordon, A. M., Huxley, A. F., & Julian, F. J. (1966). Tension development in highly stretched vertebrate muscle fibres. *Journal of Physiology, London, 184,* 143–69.

Gordon, A. M., Huxley, A. F., & Julian, F. J. (1966). The variation in isometric tension with sarcomere length in vertebrate muscle fibres. *Journal of Physiology, London, 184,* 170–92.

Armstrong, C. M., Huxley, A. F., & Julian, F. J. (1966). Oscillatory responses in frog skeletal muscle fibres. *Journal of Physiology, London, 186,* 26–7P.

Huxley, A. F., & Simmons, R. M. (1968). A capacitance–gauge tension transducer. *Journal of Physiology, London, 197,* 12P.

Huxley, A. F. (1969). Theories of muscular contraction. *Proceedings of the Royal Institution of Great Britain, 43,* 71–84.

Huxley, A. F. (1969). Theories of muscular contraction. *The Gazette, St. George's Hospital, 54,* No. 2, 5–8.

Huxley, A. F. (1970). Energetics of muscle. *Chemistry in Britain, 6,* 477–9.

Huxley, A. F., & Simmons, R. M. (1970). A quick phase in the series-elastic component of striated muscle, demonstrated in isolated fibres from the frog. *Journal of Physiology, London, 208,* 52–3.

Huxley, A. F., & Simmons, R. M. (1970). Rapid 'give' and the tension shoulder in the relaxation of frog muscle fibres. *Journal of Physiology, London, 210,* 32–3P.

Brown, L. M., Gonzalez-Serratos, H., & Huxley, A. F. (1970). Elecrtron microscopy of frog muscle fibres in extreme passive shortening. *Journal of Physiology, London, 208,* 86–8P.

Huxley, A. F. (1971). The activation of striated muscle and its mechanical response [The Croonian Lecture, 1967]. *Proceedings of the Royal Society [B], 178,* 1–27.

Huxley, A. F., & Simmons, R. M. (1971). Mechanical properties of the crossbridges of frog striated muscle. *Journal of Physiology, London, 218,* 59–60P.

Huxley, A. F., & Simmons, R. M. (1971). Proposed mechanism of force generation in striated muscle. *Nature, London, 233,* 533–8.

Huxley, A. F. (1972). Research on nerve and muscle. In *Perspectives in Membrane Biophysics,* pp. 311–17. Chicago: Gordon & Breach, The University of Chicago Press.

Huxley, A. F., & Simmons, R. M. (1972). Tension transients after quick release in rat and frog skeletal muscle. Reply to Blangé, Karemaker & Kramer. *Nature, London, 237,* 282–3.

Huxley, A. F., & Simmons, R. M. (1972). Reply to R. H. Abbott. *Nature New Biology, 239,* 186–7.

Huxley, A. F. (1973). A note suggesting that the cross-bridge attachment during muscle contraction may take place in two stages. *Proceedings of the Royal Society [B], 183,* 83–6.

Huxley, A. F., & Simmons, R. M. (1973). Mechanical transients and the origin of muscular force. *Cold Spring Harbor Symposia on Quantitative Biology, 37,* 669–83.

Ford, L. E., & Huxley, A. F. (1973). A moving coil device giving step displacements of 0.2 mm in 0.3 ms. *Journal of Physiology, London, 232,* 22P (title only).

Huxley, A. F. (1974). Muscular contraction [Review Lecture]. *Journal of Physiology, London, 243,* 1–43.

Ford, L. E., Huxley, A. F., & Simmons, R. M. (1974). Mechanism of early tension recovery after a quick release in tetanized muscle fibres. *Journal of Physiology, London, 240,* 42–3P.

Huxley, A. F. (1975). The origin of force in skeletal muscle. *Ciba Foundation Symposium, 31,* 271–90.

Huxley, A. F. (1975). New directions in muscle research. *Muscular Dystrophy Journal,* Summer 1975, 6–7.

Ford, L. E., Huxley, A. F., & Simmons, R. M. (1976). The instantaneous elasticity of frog skeletal muscle fibres. *Journal of Physiology, London, 260,* 28–9P.

Huxley, A. F. (1976). Preface in *The Motor System: Neurophysiology and Muscle Mechanisms,* ed. M. Shahani. New York, Amsterdam: Elsevier.

Huxley, A. F. (1977). Looking back on muscle. In *The Pursuit of Nature,* by A. L. Hodgkin et al., pp. 23–64. Cambridge: Cambridge University Press.

Huxley, A. F. (1977). Pure and applied research on muscle. In *Pathenogenesis of Human Muscular Dystrophies*, ed. L. P. Rawland, pp. 1–6. Amsterdam: Excerpta Medica.

Ford, L. E., Huxley, A. F., & Simmons, R. M. (1977). Tension responses to sudden length change in stimulated frog muscle fibres near slack length. *Journal of Physiology, London, 269*, 441–515.

Huxley, A. F. (1978). On arguing from one kind of muscle to another. In *Biophysical Aspects of Cardiac Muscle*, ed. M. Morad, pp. 3–23. New York: Academic Press.

Huxley, A. F., & Read, G. L. (1979). An automatic smoothing circuit for the input to digitizing equipment. *Journal of Physiology, London, 292*, 11–12P.

Huxley, A. F. (1980). *Reflections on Muscle.* Liverpool: Liverpool University Press.

Huxley, A. F. (1980). Future prospects. In *The Muscular Dystrophies, British Medical Bulletin, 36*, 199–200.

Huxley, A. F. (1980). The mechanical properties of cross-bridges and their relation to muscular contraction. *Proceedings of the International Union of Physiological Sciences, 14*, 24. (See also A. F. Huxley, 1981).

Huxley, A. F. (1980). The variety of muscle activating systems. In *Muscular Contraction: Its Regulatory Mechanisms*, ed. S. Ebashi, K. Maruyama, & M. Endo, pp. 3–18. Tokyo: Japan Scientific Society Press. New York: Springer-Verlag.

Huxley, A. F. (1980). Preface. In *Development and Specialization of Skeletal Muscle*, ed. D. F. Goldspink. S. G. B. Seminar Series No. 7. Cambridge: Cambridge University Press.

Huxley, A. F., & Lombardi, V. (1980). A sensitive force transducer with resonant frequency 50 kHz. *Journal of Physiology, London, 305*, 15–16P.

Ford, L. E., Huxley, A. F., & Simmons, R. M. (1981). Relation between stiffness and filament overlap in stimulated frog muscle fibres. *Journal of Physiology, London, 311*, 219–49.

Huxley, A. F. (1981). The mechanical properties of cross-bridges and their relation to muscle contraction. In *Advances in Physiological Sciences*, Vol. 5, ed. E. Varga et al., pp. 1–12. Budapest: Akadémiai Kiadó. Oxford: Pergamon Press.

Huxley, A. F., Lombardi, V., & Peachey, L. D. (1981). A system for fast recording of longitudinal displacement of a striated muscle fibre. *Journal of Physiology, London, 317*, 12–13P.

Huxley, A. F., Lombardi, V., & Peachey, L. D. (1981). A system for recording sarcomere longitudinal displacements in a striated muscle fibre during contraction. *Bolletino della Società Italiana di Biologia Sperimentale, 57*, 57–9.

Huxley, A. F. (1982). Skeletal muscle. *XVII Scandinavian Congress of Physiology and Pharmacology, Reykjavik, Abstracts.*

Huxley, A. F. (1984). Response to "Is stepwise shortening an artifact?" *Nature, London, 309*, 713–14.

Brown, L. M., Gonzalez-Serratos, H., & Huxley, A. F. (1984). Structural studies of the waves in striated muscle fibres shortened passively below their slack length. *Journal of Muscle Research and Cell Motility, 5,* 273–92.

Brown, L. M., Gonzalez-Serratos, H., & Huxley, A. F. (1984). Sarcomere and filament lengths in passive muscle fibres with wavy myofibrils. *Journal of Muscle Research and Cell Motility, 5,* 293–314.

Ford, L. E., Huxley, A. F., & Simmons, R. M. (1985). Tension transients during steady shortening of frog muscle fibres. *Journal of Physiology, London, 361,* 131–50.

Ford, L. E., Huxley, A. F., & Simmons, R. M. (1986). Tension transients during the rise of tetanic tension in frog muscle fibres. *Journal of Physiology, London, 372,* 595–609.

Huxley, A. F. (1986). Comments on *"Quantal mechanisms in cardiac contraction"*. *Circulation Research, 59,* 9–14.

Huxley, A. F. (1986). Discoveries on muscle: observation, theory, and experiment. *British Medical Bulletin, 293,* 115–17.

Huxley, A. F. (1986). Understanding muscle. *American Journal of Medical Genetics, 25,* 617–21.

Huxley, A. F. (1987). Studies of muscle and nerve 1957–1960 with an electron microscope provided by the Wellcome Trust at the Physiological Laboratory, University of Cambridge. *Journal of Physiology, London, 382,* 7P (title only).

Huxley, A. F. (1987). Muscle: present problems and historical background. *Pflügers Archiv, 408,* 537.

Huxley, A. F. (1988). Prefatory chapter: Muscular contraction. *Annual Review of Physiology, 50,* 1–16.

Huxley, A. F. (1990). A theoretical treatment of diffraction of light by a striated muscle fibre. *Proceedings of the Royal Society [B], 241,* 65–71.

3

Ultraslow, slow, intermediate, and fast inactivation of human sodium channels

REINHARDT RÜDEL AND BERND FAKLER

Introduction (by R. Rüdel)

After Hodgkin and Huxley (1952a,b) had described the currents underlying excitation of the squid axon, many investigators attempted to explore the mechanism of excitation in other cells such as myelinated nerve (Frankenhäuser & Huxley, 1964) and mammalian heart (Noble, 1966). I joined Trautwein's laboratory in 1965, where at the time experiments were carried out with the aim "to clamp the sodium current" in sheep Purkinje fibres (Dudel et al., 1966, Dudel & Rüdel, 1970). During a postdoctoral fellowship in Andrew Huxley's laboratory in 1968–70, I was diverted to the inward spread of activation in skeletal muscle in accordance with Huxley's more recent interests, but, back in Germany, I returned to excitation, this time to disorders of excitation in human muscle, in particular in the myotonias (Rüdel & Lehmann-Horn, 1985) and periodic paralyses (Rüdel & Ricker, 1985).

Inactivation mechanisms

In their first mathematical description of the transient sodium current, $I_{Na}(t)$, through the membrane of excitable cells, Hodgkin and Huxley (1952d; HH) postulated two independent mechanisms for the activation,

This work was supported by a grant from the DFG (Ru 138). The studies on slow and intermediate inactivation of the sodium channel were part of Bernd Fakler's M.D. thesis.

m, and inactivation, h, of the passage of ions through membranes. The equation $I_{Na}(t) = g_{Na} (E - E_{Na}) m^3h$, with g_{Na} the maximum sodium conductance, and $(E - E_{Na})$ the difference between the membrane potential and the sodium equilibrium potential, gave good fits to the experimental data, when first-order kinetics were assumed to be valid for the time dependence of m and h. In modern terms, the sodium ions are assumed to pass the membrane through channel proteins which are able to adopt conducting and nonconducting states. Transitions between these states that are mediated by three independently operating m-gates of activation and an h-gate of inactivation are compatible with the original HH model.

This model now requires extension, as several inactivation properties of the sodium channel have been described since it first appeared. Adelman and Palti (1969; squid axon), Fox (1976; frog myelinated nerve), Brismar (1977; toad myelinated nerve), Rudy (1978; squid axon), and Simoncini and Stühmer (1987; rat skeletal muscle) described inactivation processes with time constants in the range of several seconds to minutes. These "slow" or "ultraslow" inactivation processes were ascribed to an s-system which resembles the fast h-system in that it also possesses a strong dependence on the membrane potential; yet it is considered as a separate entity. Fast and slow inactivation mechanisms have in common that they do not block the passageway through the "activatable" channel when the membrane polarisation is sufficiently negative. They block it upon depolarisation.

In some of the systems investigated, the time constant of inactivation depended not only on the potential, but also on the ionic milieu of the extracellular fluid. For example, Adelman and Palti (1969) and Peganov, Khodorov and Shiskova (1973; frog myelinated nerve) reported inactivation processes that could be observed only in the presence of extracellular potassium, with time constants of some hundred milliseconds (ms; so-called intermediate inactivation; i-system). However, such a dependence could not be confirmed by Schauf, Pencek, and Davis (1976; *Myxicola* axon), Chiu (1977; myelinated nerve), or Benndorf and Nilius (1987; murine myocardial cells), even though intermediate inactivation was present. To describe this process, Chiu (1977) extended the HH model by introducing another inactivated state of the channel (second-order kinetics).

Another special feature is a "delayed" inactivation, first described by Oxford and Pooler (1975; lobster axon) and confirmed by Bezanilla and Armstrong (1977; squid axon) and Goldman and Kenyon (1982; *Myxi-*

cola axon). From the study of the gating currents, the delayed action was explained by an activation–inactivation coupling; that is, the channel can enter the state of inactivation only from the activated state (Bezanilla & Armstrong, 1977).

Experiments on the human sodium channel

In view of this complicated situation, we studied the inactivation of a human sodium channel, in order to investigate further the pathomechanisms in diseases characterised by overexcitability, such as in recessive generalized myotonia (Rüdel, Ricker, & Lehmann-Horn, 1988), or decreased excitability, such as in adynamia episodica hereditaria (Ricker et al., 1989). We chose as a preparation spherical cells of the human medulloblastoma cell line TE 671, since these are very suitable for voltage clamping in the whole-cell recording mode (Hamill et al., 1981). Details of the culturing of these cells and the voltage-clamping procedure are given in Fakler et al. (1990).

For the study of slow and ultraslow inactivation, a given cell was depolarised from the holding potential of −85 mV to a conditioning prepotential that was varied from −60 to +20 mV in 20-mV increments. For a complete assessment of the influence of the membrane potential on slow inactivation, we sometimes also incorporated hyperpolarising conditioning prepulses to −95, −115, and −135 mV in the pulse sequence. For each conditioning potential the duration of the pulse was logarithmically increased from 2 s to 6.7 min. Since the test period was short, the conditioning time was cumulative, totalling 16.3 min. After each conditioning pulse, the processes of fast and intermediate inactivation were reversed by hyperpolarisation of the membrane to −135 mV for 500 ms. Slow inactivation was then determined by measuring the maximum of the sodium current, $I_{Na, max}$, elicited by a test pulse depolarising the membrane to −20 mV for 8 ms.

In every cell investigated, the lower the conditioning potential and the longer the conditioning pulse, the smaller was $I_{Na, max}$. A plot of $I_{Na, max}$ against the duration of the conditioning pulse shows that this slow inactivation is not a monoexponential process. In the time interval studied the data could be fitted by the sum of two exponentials (Fig. 3.1). The two processes involved might be called "slow" and "ultraslow" inactivation. Their time constants, τ_s and τ_u, are in the range of 1–5 min and 1.0–1.5 h, respectively, and are dependent on the membrane potential (Table

3.1). Upon hyperpolarisation of the membrane, the effects of slow inactivation were completely reversible.

Ultraslow inactivation did not reach a steady state, even when the conditioning pulse was extended to 30 min. A total "s_∞-curve" (Simoncini & Stühmer, 1987) could not therefore be established, but we can say that slow inactivation exists over the whole potential range from −135 to +20 mV and that the inactivation curve is positioned at less negative potentials than the h_∞-curve. At the resting potential of −85 mV, already 20% of the channels seemed to be in the state of slow inactivation. We agree with Ruff, Simoncini, & Stühmer (1988) that slow inactivation may play a role in diseases with reduced resting potential of the muscle fibres, although the importance of this effect is not yet clear.

The pulse protocol for the study of intermediate inactivation was as follows: each conditioning–test pulse sequence was preceded by a hyperpolarisation of the membrane to −135 mV lasting 500 ms to render active all sodium channels capable of being activated. In a first cycle of

Fig. 3.1. Slow (s) and ultraslow (u) inactivation of human sodium channels. The maximum amplitude of the sodium current flowing during a test pulse to −20 mV is plotted against the duration of a conditioning prepulse to +20 mV. The fit made with Marquardt's algorithm yielded the following relative weights (*N*) and time constants (*τ*) for the two processes: $N_s = 0.34$, $τ_s = 56.4$ s; $N_u = 0.66$, $τ_u = 4156$ s. *Insets:* pulse protocol and set of original current records.

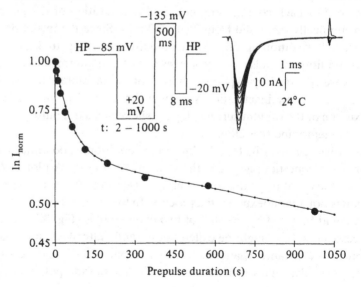

Table 3.1. *Time constants of slow (s) and ultraslow (u) inactivation (min) as a function of the prepulse potential U (mV)*

U	n	τ_s	N_s	τ_u
-135	2	0.98 ± 0.70	0.21 ± 0.08	148 ± 43
-115	2	1.85 ± 0.90	0.18 ± 0.06	153 ± 48
-95	2	2.60 ± 1.45	0.15 ± 0.05	158 ± 54
-60	2	5.94 ± 1.4	0.15 ± 0.09	117 ± 49
-40	2	2.04 ± 0.4	0.14 ± 0.11	76 ± 21
-20	2	1.6 ± 0.1	0.27 ± 0.09	53 ± 2
0	10	1.2 ± 0.5	0.28 ± 0.08	52 ± 29
$+20$	3	0.9 ± 0.0	0.39 ± 0.05	54 ± 19

Note: N_s is the fraction of slow inactivation ($N_u = 1 - N_s$). Means of n experiments $\pm$ standard deviation.

sequences, the conditioning prepotential of 8-ms duration was varied from -135 to -35 mV in 4-mV steps. After completion of the cycle the prepulse duration was doubled and the cycle repeated. Then the prepulse duration was doubled again, and so on, up to a prepulse duration of 512 ms. The result of such an experiment, illustrated in Fig. 3.2, shows that the longer the prepulse duration, the more the inactivation curve is shifted in the negative direction. With short durations, this shift arises from the fact that the prepulses are not infinitely long compared to the time constant for fast inactivation, τ_h, which is in the order of 0.5 to 50 ms, depending on the membrane potential (Hodgkin & Huxley, 1952d). The left shift of the inactivation curve seen on doubling the prepulse duration from 256 to 512 ms indicates that processes much slower than fast inactivation, but faster than slow inactivation, are involved – that is, intermediate inactivation. The processes can be separated by plotting (for one particular prepotential) the *differences* between successive current maxima against the *logarithm* of the prepulse duration. In such a time bin plot (Sigworth & Sine, 1987), a process determined by a single exponential is represented by a bell-shaped distribution, where the abscissa of the maximum corresponds to the respective time constant. Addition of several monoexponential processes is represented by superimposed "bells". The advantage of this plot is that it describes the amount of inactivation in each time bin.

Figure 3.3 illustrates such plots obtained with the same cell at conditioning prepotentials of -65 mV (A) and -45 mV (B). The two plots differ considerably in that the peaks are situated at 130 and 6 ms. This could mean that the time constant of inactivation decreases 20 fold with a potential change of 20 mV, much more than usually found. As already mentioned, we prefer the explanation that several processes of inactivation with different time constants exist, and that between -65 and -45 mV there is a shift between the relative weights, N_i, of these components. The latter possibility is confirmed by the fitting procedure. At -65 mV, three exponentials, $N_1 \exp(-t/\tau_1) + N_2 \exp(-t/\tau_2) + N_3 \exp(-t/\tau_3)$, plus a constant, were required for a perfect fit (heavy line). A fit with two exponentials (dashed line) was insufficient at long prepulse durations; the best fit with one exponential (dotted line) was insufficient at both long and short prepulse durations. More details are given by Fakler et al. (1990).

Figure 3.4 shows the time constants thus obtained as a function of the prepotential. One of these, τ_1, corresponds to fast (HH) inactivation. The other two, τ_2 and τ_3, were interpreted to represent two components of intermediate inactivation. The slower component is seen mainly at potentials negative to -60 mV. The faster component is seen negative to -40 mV. At potentials positive to -40 mV, only fast inactivation prevails.

Fig. 3.2. Voltage and time dependence of fast (h) and intermediate (i) inactivation of human sodium channels. The inactivation curves were determined with conditioning prepulses lasting 8, 16, 32, 64, 128, 256, and 512 ms. [Reproduced by permission from B. Fakler, J. P. Ruppersberg, W. Spittelmeister, & R. Rüdel (1990). *Pflüger's Archiv, 415,* 693–700.]

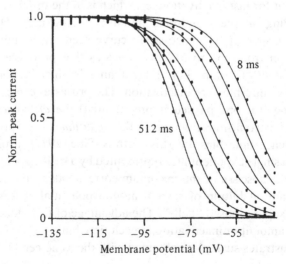

For a complete understanding of the inactivation process, the recovery from inactivation must also be investigated. For these experiments the conditioning pulse depolarised the membrane to 0 mV for 300 ms, which was sufficient to inactivate all sodium channels without turning on slow inactivation. The membrane was then repolarised to a variable negative potential for a variable duration, and then the proportion of the reprimed channels was determined with test pulses to −20 mV. Fitting sums of exponentials to the experimental data, we found again that one exponential was not sufficient for a satisfactory fit, but, in contrast to inactivation, two exponentials with time constants τ_4 and τ_5 gave good fits for all prepotentials, τ_5 being similar to τ_3, the time constant found for the slow component of intermediate inactivation (Fig. 3.4).

A recovery from inactivation characterised by two time constants cannot be explained by the model of Chiu (1977) because, according to it,

Fig. 3.3. Comparison of time dependence of inactivation for conditioning prepulse potentials of (A) −65 mV (*inset:* transient sodium currents) and (B) −45 mV. Test potential, −20 mV throughout. The shortest prepulse duration (*t*) was 3 ms; for each repeat, the duration increased by a factor of 1.33; the longest prepulse was 1.6 s. The difference between the current peaks obtained for *t* and 1.33 *t* was normalised and plotted against the prepulse duration. *Dotted line:* 1 exponential fitted to the experimental data obtained with −65 mV prepotential; *dashed line:* 2 exponentials; *heavy line:* 3 exponentials ($N_1 = 0.12$, $\tau_1 = 29$ ms, $N_2 = 0.70$, $\tau_2 = 129$ ms, $N_3 = 0.18$, $\tau_3 = 544$ ms); *thin line:* best fit to data obtained at a prepotential of −45 mV with 2 exponentials ($N_1 = 0.78$, $\tau_1 = 6$ ms, $N_2 = 0.22$, $\tau_2 = 23$ ms). [Reproduced by permission from B. Fakler, J. P. Ruppersberg, W. Spittelmeister, & R. Rüdel (1990). *Pflüger's Archiv, 415,* 693–700.]

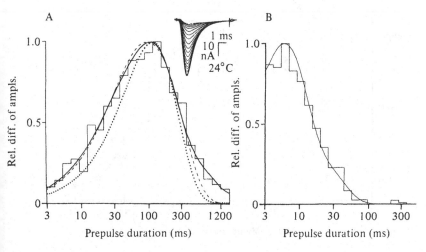

even with several steps leading to the open (=capable of being activated) state, only the rate-limiting step should show up in the tests (Meeder & Ulbricht, 1987). Since we found three time constants involved in the process of inactivation and two time constants involved in recovery, we propose the following cyclic diagram as the simplest model that describes our data:

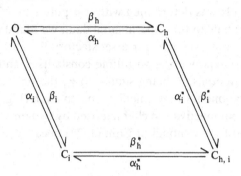

The diagram represents only inactivation, which we assume is independent of activation. It contains the state of fast inactivation, C_h, and two states of intermediate inactivation, C_i and $C_{h,i}$; slow inactivation is not incorporated. The transitions from O, the state in which the channel is

Fig. 3.4. Potential dependence of the time constants for fast (HH, τ_1) and intermediate (τ_2 & τ_3) inactivation and for the recovery from inactivation (τ_4 & τ_5).

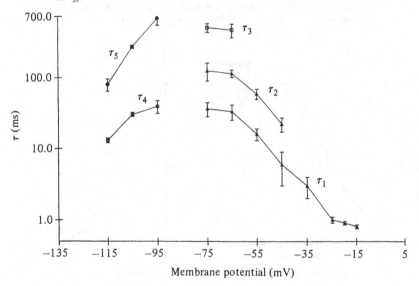

not inactivated (i.e., is activated or is capable of being activated), to the inactivation states, and the transitions between the inactivation states, are represented by a set of differential equations that contain the rate constants given in the diagram. The change in time of the state O (which corresponds to our experimentally determined current amplitudes) is derived from this set of equations as a third-order differential equation, the solution of which is the sum of three exponentials. The τ_i values of this expression are complicated functions of the rate constants. The state $C_{h,i}$ is assumed to be the last one reached during a long-lasting depolarisation. During recovery from inactivation, there are two alternatives for the reaction $C_{h,i} \rightarrow O$, which explains the occurrence of two time constants.

The interpretation of the data with the above diagram shows that inactivation is a complex process. Positive to -35 mV, fast inactivation is predominant (O$\rightarrow C_h$). With more negative membrane potentials, intermediate inactivation plays an increasing role.

Use dependence of antiarrhythmic drugs

Local anaesthetics and antiarrhythmic drugs produce effects similar to those associated with the slower components of inactivation: they reduce the sodium current and shift the inactivation curve towards more negative potentials (Hille, 1977; Meeder & Ulbricht, 1987; Wang et al., 1987). We were particularly interested in the effect of the antiarrhythmic *tocainide* because it is characterised by a pronounced "use-dependence"; that is, blockade of the sodium channels is a function of the frequency of activation. The use dependence might be the reason for the drug's efficacy in the symptomatic treatment of myotonia (Rüdel et al., 1980). The drug was quickly applied and washed out by shifting the pipette with the attached cell between parallel streams of extracellular solution, one of which contained 0.3 mM tocainide. To prevent use dependence from occurring in the presence of tocainide, conditioning pulses up to -135 mV for 500 ms were applied before each pulse.

In the presence of tocainide the inactivation curves were shifted towards more negative potentials, and, with prepulses of 512-ms duration, this shift was about three times as large as with 32-ms prepulses. Corresponding results with local anaesthetics were reported by Bean, Cohen, and Tsien (1983) and Elliott, Haydon, and Henry (1987). The experiments with this antiarrhythmic drug were essential for the detailed construction of the above model.

The state of intermediate inactivation may play a minor physiological role during a single action potential, but since the time constants of intermediate inactivation become small in the potential range positive to -35 mV, this state should be reached by an increasing number of channels during a train of repetitive activations. Within the short periods of repolarisation between action potentials, the channels do not reach the state of being activated, since in the negative potential range the time constants are large. The higher the frequency of activation, the greater the number of channels in the state of intermediate inactivation, and therefore capable of binding the drug. The hypothesis that tocainide and related drugs bind to sodium channels only when these are in the state of intermediate inactivation provides a straightforward general explanation for the use dependence of antiarrhythmic drugs.

4

The structure of the triad: local stimulation experiments then and now

CLARA FRANZINI-ARMSTRONG

Introduction

Clay and I were both finishing a period of training at the National Institutes of Health in 1964, and we wanted to go to a laboratory that was attractive to both of us. With some apprehension, and a good deal of awe, we wrote to Professor Huxley. We were surprised and delighted when a very kind reply came back. We thus came to London for a period of two years, at a time when muscle research was very exciting indeed. Fred Julian was finishing his term there, and Hugo Gonzáles-Serratos was doing his "wavy fibrils" experiments.

Clay worked downstairs, initially on crab muscle and later on length-clamp experiments, and I worked upstairs, sharing the electron microscope with Lucy Brown. I was given the task of bringing crustacean muscle in line with the sliding-filament model. As Prof (as he was called) well knew, crustacean muscles offered all sorts of possibilities to the morphologist. Bob Eisenberg showed us how to dissect crab fibres. With Prof's advice and with friendly help and encouragement from Sally Page, these turned out to be very interesting. Prof was always happy to transfer some of his knowledge of optics, and several times suggested ingenious tricks to solve questions of magnification.

This work was supported by the Muscular Dystrophy Association and NIH Grant HL 15838 to the Pennsylvania Muscle Institute. The work discussed in this manuscript is the result of collaborations with Drs. G. Nunzi, E. Varriano-Marston, D.G. Ferguson, K. Loesser, L. Castellani, K.P. Campbell, and his co-workers. I am grateful to Denah Appelt for excellent assistance.

The support that I received at the time when I had to interrupt my period of training with the arrival of our first child, as well as Richenda's example, are in good part responsible for the fact that we proceeded to have a relatively large family.

I still bear the guilt of being responsible for making Prof lose a tennis match at the annual departmental outing, but I am sure that he has long ago forgotten and forgiven.

Local stimulation experiments and membrane structure

The series of events that leads from the initial surface depolarisation of a muscle fibre to the contraction of the myofibrils [excitation–contraction (e–c) coupling] is now well established. Depolarisation of the surface membrane is carried to the fibre interior along the transverse (T-) tubules. Depolarisation of the T-tubule membrane is sensed by voltage sensors within it. Information is then transmitted, in a manner not yet fully understood, to the calcium release channels of the sarcoplasmic reticulum (SR), which briefly assume the open configuration. Calcium exits the SR, falling down a large concentration gradient, and associates with troponin C. The consequence of calcium binding to troponin C is a release of the troponin–tropomyosin inhibition of the actomyosin ATPase, and then contraction.

The local stimulation experiments of A. F. Huxley and collaborators mark the beginning of the modern era of e–c coupling research. Muscle activation was clearly a topic of widespread interest in the European scientific community of the forties and fifties, and many of the relevant questions had been posed during that period of time. The stage was set for an unusual physiologist, who had an excellent understanding of optics, an appreciation of the light microscope, the imagination to use his eyes, and the meticulous approach needed to dissect single muscle fibres and to make the special pipette tips needed for the experiments. The local stimulation experiments (A. F. Huxley, 1957a, 1974; A. F. Huxley & Straub, 1958; A. F. Huxley & Taylor, 1958; Peachey and A. F. Huxley, 1964) were designed to establish whether a structural component of the muscle fibre could be assigned the role of speeding transfer of the initial surface depolarisation to the fibre interior so that a fibre of relatively large cross-section could be activated within the relatively short period of time available. Since some evidence for a transverse continuity at the level of the Z-lines existed for some muscle fibres, the first experiments focussed on comparing the effect of depolarisation of a small patch of membrane at the level of the Z-line with that of stimulation at other levels of the sarcomere. In frog fibres a positive effect was found at the Z-line, but in crab fibres sensitive spots were located close to the A-band–I-band (A–I) junction. Luckily, electron microscopy began to contribute to the understanding of muscle structure (Robertson, 1956), and the location of triads was thereby first described (Porter & Palade, 1957).

The rest is history: a correlation was made between the positions of sensitive spots and the location of triads (A. F. Huxley & Taylor, 1958); the central element of the triad was shown to be a continuous, transversely oriented tubular network [the transverse (T-)tubules, Andersson-Cedergren, 1959; Fig. 4.1]; the lumen of the T-tubules was demonstrated to be continuous with the extracellular space (Smith, 1961; Endo, 1964;

Fig 4 1 The existence in vertebrate skeletal muscle fibres of transversely arranged invaginations of the surface membrane (T-tubules) was predicted by the results of the local stimulation experiments "If the Z membrane (distinct from the Z lines of myofibrils) contained channels whose lumen was connected to the external fluid and whose walls had a high resistance, then the potential difference across the walls would follow that across the surface membrane of the fibre" (A. F Huxley, 1957a) The micrograph is a cross-section of fish muscle showing T-tubules at the level of the Z-line ×20,000

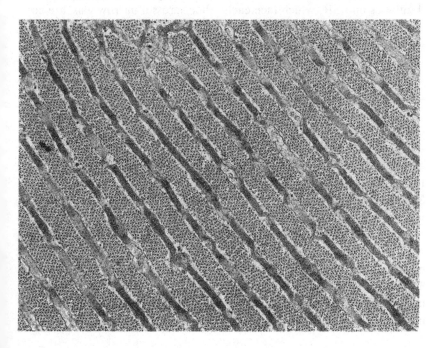

Franzini-Armstrong & Porter, 1964; H. E. Huxley, 1964; Page, 1964) and e–c coupling was off to a long run of research.

The release of calcium from the sarcoplasmic reticulum

Once the initial correlation between structure and events in the turning on of the fibre was made, the next conceptual step was immediate. The sarcoplasmic reticulum (SR) is the membrane system responsible for holding calcium in the muscle fibre (Weber, Herz, & Reiss, 1963), and it is logical to expect that release of calcium must occur from the SR, under command by the T-tubules. The first formal postulation of the role of the triads in this transmission was made by Andersson-Cedergren (1959), who wrote:

> The contact relations at the triads between the T system on the one hand and the A and I systems (now called the SR) on the other could possibly be interpreted as corresponding to a primitive synaptic relation. . . . [A]n intermediate step in the stimulation of the myofilaments involves a liberation of an exciting agent from the A and I systems of the sarcotubular system and this liberation is initiated through the stimulus propagated through the T system.

T-tubules and SR, which face each other across a narrow gap, are connected by periodically disposed proteins, forming in the junctional feet (Fig. 4.2). Physiological, biochemical, structural, and, more recently, molecular studies together have provided a good deal of information on events occurring at the T–SR junctions. I shall focus on the structural contribution to the understanding of e–c coupling.

The junctional SR cisternae forming the triads, also called *terminal cisternae,* are continuous with the rest of the SR network, the longitudinal SR, and contain a large amount of a calcium-binding protein, calsequestrin. The calsequestrin content makes these vesicles denser than those derived from the longitudinal SR (hence the term "heavy SR"), and allows them to be separated in a density gradient. Light SR vesicles, derived from the longitudinal SR, lack calsequestrin and feet and have a high ATPase content. The density of ATPase on the surface of light SR and on the free surface of heavy SR is uniformly high. In most muscle fibres, the density of ATPase in the SR membrane is $30,000–34,000/\mu m^2$ (Franzini-Armstrong & Ferguson, 1985). Muscle fibres with different rates of activity vary in their requirements for calcium uptake by the SR. Fast fibres that are capable of an extended period of

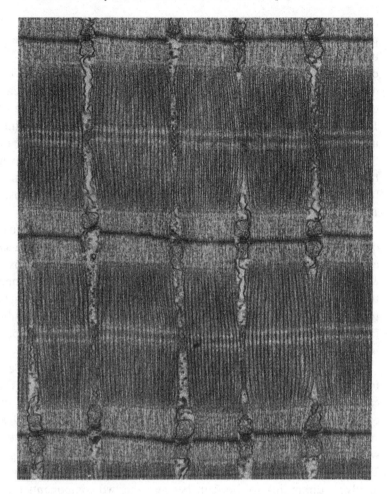

Fig. 4 2 Triads first attracted the attention of electron microscopists in 1956 (Robertson), and 1957 (Porter & Palade) The different positions of triads in frog and lizard muscle correlated well with the positions of sensitive spots in the local stimulation experiments With characteristic restraint, A. F Huxley and Taylor (1958) wrote "Both in frog and in lizard muscle the region which responds as a unit to local depolarisation is centered on the position where triad structures are formed although these positions are very different in the two animals This suggests strongly that one of the components of the triads is the structure involved" The central T-tubules are connected to the two lateral components of the triad by junctional feet, two for each junction in this view of frog muscle triads cut at right angles to the T-tubule axis ×23,000 [Reproduced by permission from C Franzini-Armstrong (1970a) *Journal of Cell Biology, 47*, 488–99]

activity need more efficient pumping than fast fibres that cannot sustain prolonged twitching. Slow fibres need the least. These requirements are met by varying either the properties of the calcium ATPase or the content of ATPase. Two isoforms of the calcium ATPase have been observed – one common to slow twitch fibres and cardiac muscle, and one found in fast twitch fibres (Brandl et al., 1986; Jorgensen et al., 1988). The total content of ATPase (number of molecules per unit volume of fibre) is varied by changing either the density of ATPase on the SR surface, or the surface density of SR in the fibre. Most fibres maintain an almost maximum packing of the ATPase in the SR membrane, but vary the surface density of SR (Ferguson & Franzini-Armstrong, 1988).

Heavy SR vesicles, derived from the terminal cisternae, contain calsequestrin, have calcium-pump ATPase in their membrane, and feet associated with their surface (Meissner, 1975; Campbell, Franzini-Armstrong, & Shamoo, 1980). Feet and ATPase occupy distinct membrane domains, the former being restricted to the area of membranè directly facing the T-tubules (or junctional SR), the latter occupying the free SR membrane, facing the myofibrils. The spatial separation between the two components is maintained in the isolated heavy SR vesicles.

Calsequestrin is an acidic protein, with a high capacity but low affinity for calcium (MacLennan & Wong, 1971); its role is to increase the SR capacity as a calcium sink. It is found in specialised regions of the endoplasmic reticulum of other cell types as well (Volpe et al., 1988). In vivo, calsequestrin is tethered to the SR membrane by thin strands, which prevent it from diffusing away into the longitudinal SR (Fig. 4.3). The association of calsequestrin with the junctional SR membrane may involve a direct link to the feet, because the two components together resist extraction of the membrane by detergents (Caswell & Brunschwig, 1984).

The structure of the junction

The junctional feet are disposed in an orderly array. In fibres of vertebrates, the junctional gap is occupied by two or more rows of feet, at a centre-to-centre distance of 28 nm (Fig. 4.2). In en face views of the junction, individual feet have an approximately diamond (or square) shape and are obliquely arranged along the rows, so that they touch each other approximately at the corners (Fig. 4.4). The structure of feet, at a limited level of resolution, can be seen by rotary shadowing of the surface of isolated heavy SR vesicles (Ferguson, Schwartz, & Franzini-Armstrong, 1984; Fig. 4.5). Each foot is composed of four, apparently identical subunits, symmetrically disposed around a central depression ("quatrefoils", or four-leaf clover). Adjacent feet are associated corner to corner but with a twist, so that the diagonal of each foot is slightly rotated relative to the axis of the parallel rows. The specific disposition is maintained in isolated heavy SR, and thus it may not depend on interaction with components of the T-tubules.

Junctions between special SR elements and the surface membrane or its invagination are a constant component of all muscle fibres, including cardiac and smooth muscle. Junctional feet are present in all these junctions. Interestingly, feet in muscles of invertebrates that greatly diverge from the vertebrates on the phylogenetic tree (Mollusca, Arthropoda) have the same quatrefoil structure, but differ in disposition from those of vertebrates (compare Figs. 4.4–4.5 with 4.6–4.7).

Fig. 4.3. Calsequestrin has the important function of increasing the SR capacity for calcium. It is anchored to the SR membrane so that it remains in the triad. Within the split, free SR membrane are particles showing the extremely dense disposition of the calcium ATPase. Deep-etch view of a fish muscle. ×84,000. [Reproduced by permission from C. Franzini-Armstrong, L. J. Kenney, & E. Varriano-Marston (1987). *Journal of Cell Biology, 105,* 49–56.]

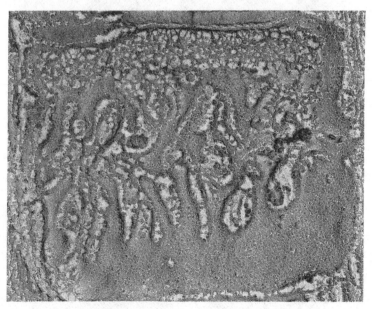

The junctional feet are responsible for holding SR and T-tubules at a small fixed distance from each other. A series of recent breakthrough experiments showed that they also play a key role in e–c coupling, since they are the sites for calcium release from the SR. First, it was shown that the heavy SR contains an intrinsic component of large molecular

Figs. 4.4–4.7. Views of feet in thin sections grazing to the junction (Fig. 4.4, fish), and after rotary shadowing of a "heavy SR" vesicle (Fig. 4.5, guinea pig). Both from twitch fibres. The feet have an orderly arrangement in two parallel rows. Each foot is composed of four equal subunits (quatrefoil). Figures 4.6 and 4.7 are as Figs. 4.4 and 4.5, but in a scorpion and a grasshopper, respectively. The quatrefoil structure of feet is the same as for vertebrate muscle, but the disposition is different. Figures 4.4 and 4.6, ×66,000; Fig. 4.5, ×108,000; Fig. 4.7, ×54,000. See Loesser et al., 1992.

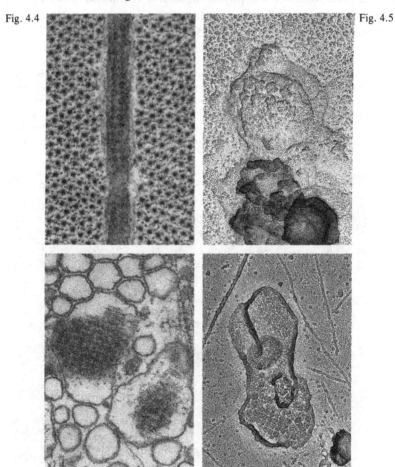

Fig. 4.4

Fig. 4.5

Fig. 4.6

Fig. 4.7

weight which can be identified as a subunit of the junctional feet (Kawamoto et al., 1986; Volpe, Gutweniger, & Montecucco, 1987). Concomitantly, it was shown that a channel formed by incorporating heavy SR vesicles in a lipid bilayer has the same pharmacological properties as the rapid calcium release from the vesicles (Meissner, Darling, & Eveleth, 1986; Smith, Coronado, & Meissner, 1986). This channel has been called the SR calcium release channel, and it is thought to be the site through which calcium exits from the SR during e–c coupling. Finally, a large protein was isolated, identified as the feet, and shown to form the calcium release channel (Inui, Saito, & Fleischer, 1987; Lai et al., 1988; Smith et al., 1988). The protein is variously called spanning protein, foot protein, junctional complex, and ryanodine receptor. The last name results from the fact that the protein has a high affinity for ryanodine, a plant alkaloid that blocks the channel in an open state with lower conductance (Meissner, 1986); this produces calcium release from heavy SR (Fleischer et al., 1985) and a contracture in muscle fibres (Sutko, Ito, & Kenyon, 1985). Ryanodine has been a very important tool in the isolation of the foot protein. Identity of the SR release channel with the junctional foot is direct, owing to the unique size and shape of this large molecule (Figs. 4.8 & 4.9). The identification has important consequences for e–c coupling. We now know that the calcium release sites are located in the junctional SR membrane immediately adjacent to either the T-tubules or surface membrane, and thus in the best location for receiving the coupling signal. No other sites of the SR have feet associated with them.

The foot protein has two domains: a large hydrophilic domain that spans the gap between SR and T-tubules (the foot), and a smaller domain which is located in the junctional SR membrane (Figs. 4.8–4.10). The latter portion spans the entire membrane thickness, as may be expected from its function as a channel (Kawamoto, Brunschwig, & Caswell, 1988). The internal structure of the foot protein has been determined with a fair degree of resolution (Wagenknecht et al., 1989), and the single polypeptide constituting it has been sequenced (Takeshima et al., 1989). In agreement with the structural observations, the sequence indicates a large hydrophilic segment (the cytoplasmic domain) and a shorter hydrophobic one (the channel), with some similarity to a subunit of the acetylcholine receptor. It is likely that the entire foot, composed of four identical polypeptides with a combined molecular weight of approximately 2,200 kilodaltons (kD), represents a single channel. This makes it the largest channel known to date. The uniqueness of the SR calcium release channel resides in its ex-

Figures 4.8 and 4.9. The isolated foot protein (Fig. 4.8; see Block et al., 1988) has the same quatrefoil shape as the foot adhering to the SR membrane (Fig. 4.9), but it has a central bump in place of a central depression. That is the portion of the molecule that was originally located within the SR membrane (see Fig. 4.10). Rabbit. Figure 4.8, ×225,000; Fig. 4.9, ×108,000. [Figure 4.9 reproduced by permission from B. A. Block, T. Imagawa, K. P. Campbell, & C. Franzini-Armstrong (1988). *Journal of Cell Biology, 107,* 2587–600.]

Fig. 4.8   Fig. 4.9

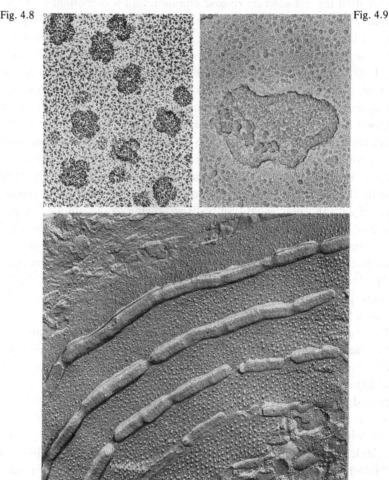

Fig. 4.10

Figure 4.10. Freeze-fracture splitting along the junctional SR membrane. The location of the intramembranous portion of each foot (the calcium release channel) is marked by a small bump. Fish muscle. ×39,000. [Reproduced by permission from B. A. Block, T. Imagawa, K. P. Campbell, & C. Franzini-Armstrong (1988). *Journal of Cell Biology, 107,* 2587–600.]

tremely large cytoplasmic domain. This is the component that is responsible for its function in coupling SR either to T-tubules (in those fibres that have them), or to the surface membrane (in fibres that have no or few T-tubules). It is likely that this is the portion of the molecule responsible for translating a T-tubule signal into gating of the release channel (or intramembranous domain of the molecule).

The next problem concerns the nature of the T-tubule signal; to explain it we need to consider the T-tubule component of the T–SR junction. In vertebrate skeletal muscle fibres, there is some evidence for the interaction between the SR feet and a specific component of the T-tubules: (1) Junctional SR membrane, presumably with attached feet, will form junctions with T-tubules in vitro (Corbett et al., 1985). (2) After freeze-fracture the junctional portion of the T-tubule membrane is occupied by large particles. In the fibres of a fish, these particles have an orderly disposition (Figs. 4.11 & 4.12). The repeating component of the array is a group of four particles, the jT tetrad. jT tetrads are arranged along rows parallel to those of the underlying feet, and at a spacing such that they are located over alternate feet (compare Figs. 4.10 & 4.11). The four components of each tetrad are located immediately above the centres of the four subunits of the underlying junctional foot, and one must assume that some link exists between the two components of the junction. Junctional tetrads have been observed in frog slow fibres as well (Franzini-Armstrong, 1984).

The components of the jT tetrads have not yet been identified. Reasons for expecting them to be modified calcium channels, the so-called dihydropyridine receptors (DHPR), are given in Block et al. (1988), and Caswell and Brandt (1989). The structure of the isolated dihydropyridine receptor is shown in Fig. 4.13 (Leung et al., 1988). The alpha subunit of the DHPR has strong homologies with the voltage-gated sodium channel (Tanabe et al., 1987).

From the above structural information, we deduce that the triad is a junction where components of one membrane, the jT tetrads, are associated with intrinsic components of the other membrane (the feet), and may be capable of a direct interaction.

The mechanism of excitation–contraction coupling

Hypotheses of e–c coupling must take into account the structures illustrated above. Calcium indicators have been effectively used to learn a good deal about the properties of calcium release during e–c coupling.

Figures 4.11 and 4.12. Freeze fracture along T-tubules showing periodic disposition of jT tetrads. These groups of T-tubule proteins are associated with alternate feet in this fish muscle. Figure 4.11, ×30,000; Fig. 4.12, ×79,800. [Fig. 4.11, see Block et al., 1988; Fig. 4.12 reproduced by permission from B. A. Block, T. Imagawa, K. P. Campbell, & C. Franzini-Armstrong (1988). *Journal of Cell Biology, 107,* 2587–600.]

Fig. 4.11

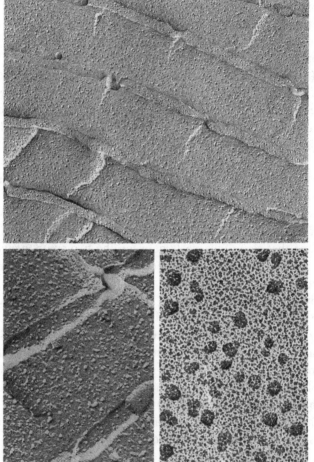

Fig. 4.12

Fig. 4.13

Figure 4.13. The purified dihydropyridine receptor has an ovoidal shape, and two visible subunits. ×225,000. [Reproduced by permission from A. T. Leung, T. Imagawa, B. Block, C. Franzini-Armstrong, & K. P. Campbell (1987). *Journal of Biological Chemistry, 263,* 994–1001.]

Unexpectedly, it was found that the rate of calcium release has a very early peak, and then declines to a lower value even with prolonged depolarisation (Baylor, Chandler, & Marshall, 1983; Meltzer, Rios, & Schneider, 1984, 1986). Whether this is an intrinsic property of the release channel or a feedback of the increased calcium concentration (Baylor & Hollingworth, 1988) is not yet known. Properties of the isolated calcium release channel and of the in situ triad are similar but not identical. Further comparisons between the two may clarify how the channel is affected by its junction to the T-tubules (S. Baylor, personal communication).

It is now generally accepted that one of the early events in e–c coupling is the internal rearrangement of a voltage sensor in T-tubules (or surface membrane), which is detected as the so-called charge movement (Schneider & Chandler, 1973). Several hypotheses have been proposed over the years to connect charge movement to SR calcium release. The idea of direct, possibly gated, current flow between the two membrane systems was abandoned a long time ago. Among the most immediate difficulties of such an electrotonic transmission at the triad are the impedance mismatch resulting from the much larger size of SR relative to T-tubules, and the high potassium permeability of the SR which renders it unable to maintain a steady-state membrane potential. A second possibility is the so-called calcium-activated calcium release hypothesis, based on the fact that under certain conditions an elevated calcium concentration induces rapid release of calcium from the SR (Endo, Tanaka, & Ogawa, 1970). This is also not likely to be the primary event in skeletal muscle, since its calcium current is slower than e–c coupling, and the fibres can contract in the absence of extracellular calcium. It cannot be excluded at this point that calcium-activated calcium release may play a role as a secondary effect in e–c coupling. Finally, it has been proposed that the T-system and SR may interact by means of long-range molecular interactions (Schneider & Chandler, 1973). The direct apposition of T-tubule components and junctional feet which is observed by electron microscopy makes this hypothesis particularly attractive.

It has been proposed that DHPRs of transverse tubules may be modified calcium channels that act as the voltage sensors of e–c coupling (Rios & Brum, 1987). This possibility is greatly strengthened by the finding that a genetic defect resulting in the lack of the DHPR alpha subunit results in a specific interruption of e–c coupling (Beam, Knudson, & Powell, 1986; Tanabe et al., 1988; Knudson et al., 1989). Thus two major molecules involved in e–c coupling are identified: the DHPR (possibly representing the voltage sensor) and the calcium release channel. It remains to be demonstrated whether they can (directly or indirectly) interact with each other. Unfortunately, this coupling has been very difficult to identify biochemically (see Caswell & Brandt, 1989), and its function is still a matter of debate.

Other interesting questions remain to be solved, for example: Why is the e–c coupling delay relatively long (Vergara & Delay, 1986)? Does the apparent interaction between calsequestrin and the feet have any functional significance in e–c coupling? Does inosine triphosphate (IP_3) act on the e–c coupling calcium release channel (the foot) or on some other receptor in skeletal muscle?

How much of what we know about transmission at the triad could have been predicted by the results of the local stimulation experiments, and how relevant are those results today? Much is actually hidden in those very careful observations. The presence of a voltage sensor in the T-tubule is a direct requirement of the strict dependence of e–c coupling on T-tubule membrane voltage. The correlation between strength of depolarisation and extent of the resultant contraction indicates some very direct functional link between T and SR, one that requires a specific functional junction between the two systems. Lack of longitudinal spread is also an indication that e–c coupling is not mediated by an initial SR depolarisation. Finally, the graded nature of the effect should rule out the calcium-activated calcium release mechanism (Endo, Tanaka, & Ogawa, 1970; Ford & Podolsky, 1970), which tends to feed upon itself and to produce an all-or-none response.

Investigators interested in the possibility of a chemical transmission step in coupling at the triad should bear in mind that in order to keep the contraction from spreading away from the depolarisation T-tubule segment, the hypothetical transmitter has to be destroyed very rapidly.

5

The calcium-induced calcium release mechanism in skeletal muscle and its modification by drugs

MAKOTO ENDO

Introduction

I joined Andrew Huxley's laboratory in the autumn of 1962, coming from H. Kumagai's laboratory in Tokyo. Contact had been made by Setsuro Ebashi, one of the senior students of H. Kumagai, who visited Andrew Huxley for the first time in 1960 on his return to Tokyo from Rockefeller University, where he had just established the essential part of his calcium theory, and from whom I had learnt muscle biochemistry and physiology. Initially, Prof and I had almost decided to start examining the influence of extracellular calcium on depolarisation–contraction coupling, but during my initial training in preparing single muscle fibres H. C. Lüttgau visited us on his way to Cambridge and disclosed that he had already done almost everything I intended to do. Then Prof suggested an alternative project – to examine whether the lumen of the T-tubule is continuous with the extracellular space by the use of fluorescent dyes, to which I devoted the entire one year and eight months of my stay in London. After coming back to Tokyo, I intended to proceed to the next step – to T-tubule depolarisation (i.e., the mechanism of calcium release from the sarcoplasmic reticulum). Although I have not yet been able to elucidate the physiological mechanism of calcium release, I have investigated problems related to the mechanism, as is described in this chapter.

Of course, I learnt very much from Prof in London, but I also received a great deal from him after I came back to Tokyo.

Professor Huxley came to Japan for the first time on September 1, 1965 to attend the XXIII International Congress of Physiological Sciences held in Tokyo, and stayed in Japan for about a month. Waseda University, one of the leading private universities in Japan, asked him to deliver a lecture to commemorate the establishment of their Department of Biophysics, and he accepted. His lecture, entitled "Physics of Nerve Conduction," was scheduled for Monday, September 27, 1965. Since the majority of the audience was expected to be general science students, not necessarily in physiology or in biophysics, his lecture was to be translated into Japanese sentence by sentence by an interpreter to facilitate the understanding of the audience. Speaking and listening to English were much more difficult for Japanese students 25 years ago than at present.

Having spent about two weeks in Japan, and having begun to learn Japanese, in which he was very much interested, Prof began to wonder if he could give his lecture in Japanese. This seemed a good idea to him because a lecture interpreted sentence by sentence could well be dull. The people at Waseda University were at first reluctant to accept the idea, apparently because they didn't believe in Prof's Japanese, but Prof was very confident because he had had a similar experience in the Soviet Union previously. After a stay in the Soviet Union for only a few weeks, and having started to learn Russian during his stay, he had successfully delivered his lecture in Russian! It was of course quite clear to him that the task would demand tremendous effort, but he liked such challenging tasks, and probably after a period of relaxation of the post-Congress private tour in Kyoto and Nara, he might well have been longing for his usual active life of concentration. In any case, his firm will finally persuaded the Waseda people to accept his proposal, although apparently with ill grace.

The English text of his lecture was handed to me on Saturday, September 25, two days before his lecture. I immediately started translating it into Japanese in a lecture style and wrote it in Roman characters. We did not have much time, so Professor Ebashi and Dr. Ohtsuki helped me with the translation in the later stages. The typescript of the text typed out by our departmental secretary was returned to Prof page by page. What Prof then did was to check the Japanese text, consulting a Japanese–English dictionary he had bought in Maruzen Bookstore several days before, partly in order to confirm and practise the pronunciation, but, what was more important, to understand what he was saying in every sentence, which is certainly essential for delivering a lecture successfully. It should be appreciated that with such a severe time limitation, this was no easy task, even for such a person as Prof. Japanese is a language entirely different from English grammatically as well as phonetically, and Prof had started learning the language only a few weeks before; and even during this period the time available for learning was very limited owing to his busy schedule. At supper time on Sunday evening, the night before his lecture, he still had a great deal of work left to do, and the prospects appeared not very good, and Prof's eyes looked bloodshot. At discussion during the supper, however, we realised that his original text was too long if spoken in Japanese, so we reduced the length substantially. His rapid progress in Japanese also speeded up his job, and when he went back to his hotel at midnight, Prof had recovered his usual calm and confident appearance. After finishing the remaining preparation and final rehearsal on Monday morning, Prof delivered a very impressive lecture.

There are pieces of evidence to indicate that Prof completely understood his Japanese text. In translating his text in a hurry, I made one careless mistake; whereas he had written "osmotic pressure is high", I somehow wrote it down in Japanese as "osmotic pressure is *low*". Prof pinpointed the error and came to my place, where I was still working on the translation. He was very polite, however, and said, "This sounds very odd, but probably you express it in Japanese in this way?"

His lecture was given with perfect pronunciation, although in a foreign accent, which was of course inevitable, but more importantly with a very appropriate pause between phrases, which clearly indicated that he understood what he was saying. The audience enjoyed his lecture very much, and when he came to a joke which he had inserted in the text, everybody laughed. The text of the lecture was subsequently published (A. F. Huxley, 1965).

We were greatly impressed with Prof's unbelievable speed of understanding of entirely new things, his immense intellectual appetite, his enormous concentration, his high fighting spirit, and his tenacity. I was very happy and much honoured to be able to support this great achievement, and in doing so I learned quite a lot.

The release of calcium from the sarcoplasmic reticulum

Among the series of processes of excitation–contraction (e–c) coupling of skeletal muscle, how the depolarisation of the T-tubule membrane causes

Ca^{2+} to be released from the sarcoplasmic reticulum (SR) has been the least understood step, since the establishment of the role of Ca^{2+} in the coupling process (Ebashi & Endo, 1968). Our strategy of approaching this problem has been to examine, using skinned fibres, what kind of stimuli, when directly applied to the SR, could cause a release of Ca^{2+}, in the hope that among effective stimuli we might encounter the one that physiologically effects signal transduction from the T-tubule to the SR. After starting studies along this line, we soon found that Ca^{2+} itself can cause a release of calcium from the SR; we called the phenomenon "calcium-induced calcium release" (Endo, Tanaka, & Ebashi, 1968; Endo, Tanaka, & Ogawa, 1970). At about the same time, Ford and Podolsky (1968, 1970) also found the same phenomenon independently. Although it was conceivable, at the beginning, that the mechanism of calcium-induced calcium release is included in the physiological T-tubule–SR coupling processes (Ford & Podolsky, 1972b), all the properties of the calcium-induced calcium release so far elucidated militate against its physiological role; in contrast, its pathophysiological role in malignant hyperthermia and its pharmacological role in the caffeine contracture have clearly been demonstrated (cf. Endo, 1977, 1985). However, recent biochemical studies in this field suggest that the calcium release itself may also act as a physiological calcium release channel in an entirely different mode of operation, a quite unexpected significance of the calcium-induced calcium release mechanism. In this chapter, I summarise the physiological and pharmacological properties of the calcium-induced calcium release mechanism and discuss the relation between the calcium-induced calcium release channel and the physiological calcium release channel.

Some methodological considerations in examining properties of calcium-induced calcium release

Two feedback effects should be considered in order to determine accurately the magnitude of activation of calcium-induced calcium release: First, since Ca^{2+} applied to the SR activates not only the calcium-induced calcium release mechanism but also the calcium pump protein, the apparent amount of calcium released by Ca^{2+} is not necessarily the result solely of calcium-induced calcium release. In other words, if the ambient Ca^{2+} concentration is elevated as a result of calcium release, it stimulates the calcium pump, and a part of calcium released may be taken up again by the SR. To avoid this negative feedback effect, calcium release should be determined in the absence of calcium pump activity; this can easily be achieved in skinned fibres by the withdrawal of ATP from the medium. Second, since Ca^{2+} released by the action of calcium applied externally could further activate calcium-induced calcium release and, therefore, amplify the amount of calcium released, this positive feedback loop should be cut off when the magnitude of the primary calcium-releasing effect of Ca^{2+} is to be determined. This can be accomplished by the use of a high concentration of calcium buffer, which effectively binds the Ca^{2+} released and reduces the resulting change in the free Ca^{2+} concentration to a minimum. The amount of calcium released must then be measured by determining the difference in the amount of calcium remaining in the SR with and without giving the calcium stimulus, because the concentration of Ca^{2+} in the ambient solution is intentionally made insensitive to calcium release in this type of experiment.

The actual procedures adopted in determining the relationship between Ca^{2+} concentration and the magnitude of activation of calcium-induced calcium release were essentially as described in Endo and Iino (1988): in mechanically or chemically (saponin 50 μg/ml, 30 min) skinned single skeletal muscle fibres or small bundles (about 100 μm in diameter), isometric tension was monitored. The following three-step procedures were repeated, with conditions varied in step 2:

Step 0 (preparatory step at the very beginning) – emptying the SR: A high concentration of a calcium-releasing agent (such as 25 mM or a higher concentration of caffeine) was applied.

Step 1 – loading the SR with calcium: The fibres were incubated with a fixed concentration of free calcium (typically pCa 6.5, highly buffered with 10 mM total EGTA*) in the presence of MgATP, and the calcium pump was allowed to accumulate calcium into the lumen of the SR for a fixed period of time (typically 3 min at 2°C).

Step 2 – releasing calcium by activation of calcium-induced calcium release: After MgATP was withdrawn to stop the action of the calcium pump, various concentrations of Ca^{2+}, highly buffered with 10 mM total EGTA, were applied for various periods of time, and calcium in the SR lumen was allowed to extrude through the activated calcium-induced calcium release channel. At the end of the period, calcium release was quickly terminated by closing the calcium-induced calcium release channel (by withdrawing free Ca^{2+} and simultaneously

---

*Ethylene glycol bis(β-aminoethyl ether)N,N'-tetraacetic acid.

applying a high concentration of inhibitors of calcium-induced calcium release, e.g., 10 mM magnesium ions plus 10 mM procaine).

Step 3 – assaying the remaining calcium in the SR: After reintroducing MgATP, all the calcium in the SR was released by applying a high concentration of caffeine, and the amount released was assayed by determining the magnitude of the resulting contracture of the skinned fibres themselves.

Since Step 3 empties the SR, the next series could start from Step 1, with Step 0 skipped. By altering the duration of application of Ca^{2+} at any fixed concentration at Step 2, one could obtain the time course of calcium release at the Ca^{2+} concentration. As shown in Fig. 5.1, the amount of calcium remaining in the SR follows an exponential time course, the rate of which is a function of Ca^{2+} concentration. This exponential time course could be interpreted as a result of the steady opening of calcium channels, the probability of opening being determined by the Ca^{2+} concentration, which causes efflux of Ca^{2+} from the SR in proportion to the concentration difference across the SR membrane.

Some properties of calcium-induced calcium release

The exponential time courses in Fig. 5.1 do not indicate inactivation of the calcium-induced calcium release, although Fabiato reported a rapid

Fig. 5.1. Relation between duration (t, min) of Ca^{2+} stimulation and logarithm of relative amount of calcium remaining in the SR. Each point and vertical line shows the mean ± SE ($n = 4$). [Reproduced by permission from T. Ohta, M. Endo, K. Nakano, Y. Morohoshi, K. Wanikawa, & A. Ohga (1989). *American Journal of Physiology, 256*, C358–67.]

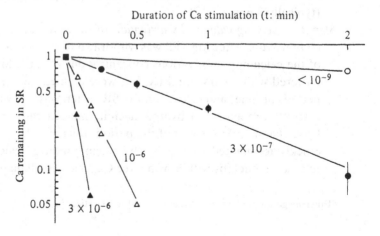

inactivation in his "calcium-induced calcium release" in cardiac muscle (Fabiato, 1985). Since the rate of calcium application in our skinned fibre experiments is limited by diffusion through 100 μm of fibre space, some initial rapid inactivation (partial and quickly completed) could not be excluded from the results such as shown in Fig. 5.1. However, the fact that the values extrapolated back to zero time from the exponential decay almost coincide with the values obtained without the addition of calcium-releasing stimuli, as shown in Fig. 5.1, indicates that the size of such a rapidly inactivating component must be a minor one, if any, at least in skeletal muscle, because Ca^{2+} should have quickly enough reached at least the SR at the surface region of the fibre.

The rate of exponential decay (as shown in Fig. 5.1) is plotted against free Ca^{2+} concentration in Fig. 5.2. As shown in the figure, in the absence of Mg^{2+} and of sensitising agents such as caffeine, a level of Ca^{2+} in the micromolar range is required to activate the release mechanism. At the foot of the rising phase, the rate increases in proportion to the square of free Ca^{2+} concentration, suggesting that two calcium ions are required to open the calcium channel. At higher concentrations of Ca^{2+} (above 100 μM), the rate decreases with increasing Ca^{2+} concentration, indicating an additional inhibitory effect of Ca^{2+} on the release mechanism.

The calcium-induced calcium release mechanism is enhanced by ade-

Fig. 5.2. Dependence of the rate of calcium release from the SR on pCa. The straight line has a slope of 2. Skinned fibres from human muscles at 20°C, in the absence of Mg^{2+}. (Modified from M. Endo et al. 1983.) Results of amphibian fibres were similar.

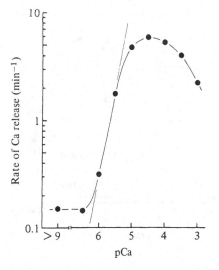

nine compounds (Endo, Kakuta, & Kitazawa, 1981) and xanthine derivatives (cf. Endo, 1985). Whereas adenine and its derivatives increase the rate of calcium-induced calcium release without altering calcium sensitivity, xanthine compounds increase the sensitivity of the calcium-induced calcium release mechanism to Ca^{2+} as well as the rate of calcium release at the optimum Ca^{2+} concentration (Fig. 5.3). ATP is the strongest agent in enhancing release among adenine compounds. A nonhydrolysable analogue of ATP, AMPPCP (β,γ-methyleneadenosine 5′-triphosphate), is also very effective; ADP and AMP are less effective in this order, and adenosine and adenine are weaker agonists in this sense. Among xanthine compounds, caffeine (1,3,7-trimethylxanthine) is the most widely used agent, and its well-known contracture-inducing effect is attributable to this enhancing effect on the calcium-induced calcium release mechanism. Roughly speaking, the calcium-induced calcium release mechanism is so sensitised by a contracture-inducing concentration of caffeine that even at the resting level of cytoplasmic free Ca^{2+} concentration, the calcium release rate overwhelms the calcium uptake rate of the calcium pump, and hence a contracture is produced. It has been reported that 1,7-dimethylxanthine is the strongest of this group, but all

Fig. 5.3. Potentiating effect of 10 mM AMP and 50 mM caffeine on Ca-induced Ca release in skinned fibres of *Xenopus* fast muscle at 2°C. Rates were determined as described in Fig. 5.1, and are plotted against Ca^{2+} ion concentrations. The Ca^{2+} stimuli were given in the absence of Mg^{2+}. (Modified from M. Endo 1981.)

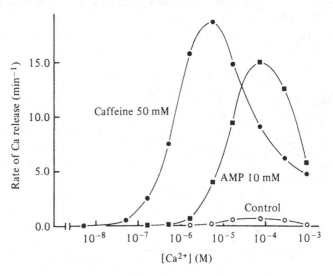

the xanthine compounds except 9-methyl derivatives are effective (Rousseau et al., 1988). In living cells, however, the apparent effects of the compounds are influenced by the penetrability of each compound through the cell membrane.

There are at least two types of inhibitors of calcium-induced calcium release. Procaine, tetracaine, and some other local anaesthetics decrease the rates of calcium-induced calcium release (Ford & Podolsky, 1972a; Thorens & Endo, 1975) without appreciably altering the calcium sensitivity. Not all the local anaesthetics have this action, and indeed some such as dibucaine and lidocaine do enhance, slightly but definitely, the calcium-induced calcium release (Yagi & Endo, 1980). Mg^{2+}, on the other hand, decreases the calcium sensitivity of the release mechanism, probably competing at the activating site with Ca^{2+}, but Mg^{2+} also decreases the maximum rate of calcium release at an optimal Ca^{2+} concentration (Fig. 5.4).

Physiological significance of calcium-induced calcium release

It is sometimes argued that because calcium-induced calcium release produces a positive feedback loop and because the physiological calcium

Fig. 5.4. Effects of Mg^{2+} and procaine on the calcium-induced calcium release potentiated by caffeine. Plots as in Fig. 5.3. Caffeine was used to make inhibitory effects clearer, but essentially the same inhibition was obtained in the absence of caffeine. [Reproduced by permission from M. Endo (1981). In *The Mechanism of Gated Calcium Transport Across Biological Membranes*, ed. S. T. Onishi & M. Endo, pp 257–64. New York: Academic Press.]

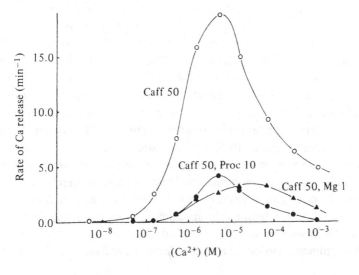

release occurs not in an all-or-none but in a graded manner, the physiological calcium release cannot be induced by calcium. However, this is not a sound argument because in living muscle cells the calcium pump, which produces a negative feedback loop, is also operating, and therefore, even in the presence of an active calcium-induced calcium release the overall feedback effect could well be negative rather than positive.

In the presence of physiological concentrations of Mg^{2+} and ATP, the level of free Ca^{2+} required to activate the calcium-induced calcium release mechanism as determined in skinned fibres is several micromolar or above, which seems too high for the mechanism to be important physiologically, unless the location of the release channels is compartmentalised and separated from other parts of the cytoplasm. The evidence enumerated in the following three paragraphs speaks further against the possibility of direct involvement of the calcium-induced calcium release mechanism in the physiological calcium release process during e–c coupling.

1. Although the maximum rate of calcium-induced calcium release as activated by, for example, a high concentration of caffeine could be quite high, the maximal rate of calcium release induced by calcium ion itself (30 μM) determined in skinned fibres under "physiological" conditions was more than an order of magnitude lower than the reported rates of physiological calcium release (Endo, 1984). Thus, Ca^{2+} cannot be the sole mediator of physiological calcium release, unless skinned fibres have lost a hypothetical substance that potentiates calcium-induced calcium release.

2. If calcium-induced calcium release is operating as the main mechanism during the physiological release of calcium, inhibitors of the calcium-induced calcium release should inhibit the physiological release as well. However, procaine (which inhibits calcium-induced calcium release, as evidenced by the inhibition of caffeine contractures) did not inhibit contractions evoked by cell membrane depolarisation during voltage clamp or during an application of a solution containing an elevated potassium concentration (Heistracher & Hunt, 1969; Thorens & Endo, 1975). Similarly, an inhibition of calcium-induced calcium release by adenine did not depress the physiological calcium release. Adenine by itself enhances calcium-induced calcium release as mentioned above, but its effect is much weaker than that of ATP, and therefore, in the presence of ATP, adenine inhibits calcium-induced calcium release, because the strong enhancing effect of ATP is replaced by the much weaker enhancing effect of adenine (Ishizuka & Endo, 1983). The effect

of adenine on intact fibres is very similar to that of procaine. Caffeine contractures were clearly inhibited by adenine as expected (Fig. 5.5A), but twitches or potassium contractures were not at all inhibited but rather potentiated. Furthermore, when twitches were potentiated by caffeine, adenine depressed the caffeine potentiation, but when twitches were potentiated by nitrate replacing chloride, adenine did not depress the potentiation at all (Fig. 5.5B; Ishizuka, Iijima, & Endo, 1983). These pharmacological studies clearly demonstrate that the calcium-induced calcium release mechanism does not operate during physiological calcium release unless it is potentiated by caffeine.

3. More recently, Baylor and Hollingworth (1988) demonstrated that even when a large buffering concentration of calcium chelator, Fura-2, was injected into skeletal muscle fibres, calcium release by action potentials was not inhibited but potentiated, although the calcium mediator hypothesis – that is, the hypothesis that calcium-induced calcium release plays the major role in e–c coupling – predicts an inhibition.

Pathophysiological significance of calcium-induced calcium release: malignant hyperthermia

Although the calcium-induced calcium release mechanism does not play any important role in physiological contraction as discussed above, its enhancement appears to cause a disease known as malignant hyperthermia (Denborough & Lovell, 1960; Gronert, 1980). This disease is a hereditary disorder in which patients treated with inhalation anaesthetics, especially halothane, develop a very high fever (over 40°C), usually with generalised muscle rigidity. Owing to metabolic disturbances secondary to the high fever and the resulting vicious circle, the mortality rate reaches 70% unless the condition is properly treated.

The conspicuous muscle rigidity led to an examination of the muscles of these patients. These muscles were demonstrated to be more sensitive to caffeine than normal muscle (Kalow et al., 1970). This finding, together with the fact that procaine could be effective in treating the disease, suggested that malignant hyperthermia is a disorder of the calcium-induced calcium release mechanism. Indeed, we showed (Endo et al., 1983) that the calcium-induced calcium release in the muscles of these patients has a higher calcium sensitivity than that in normal muscle, and the maximal rate of calcium release at an optimum Ca^{2+} concentration is also higher in the affected muscle than in normal muscle. We also demonstrated that halothane and all the other inhalation anaes-

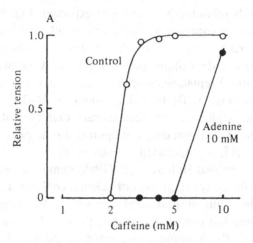

Fig. 5.5.(A) Effect of adenine on caffeine contracture in single muscle fibres. Dose–response curves at steady state. Various concentrations of caffeine were applied in the absence (control) or presence of 10 mM adenine. In the latter, a 30-min preincubation was used. The peak tension of the contracture is plotted relative to that of the maximum response in the absence of adenine. (B) Effect of adenine on twitch under various conditions: (a) effect on normal twitch; (b) effect on caffeine-potentiated twitch; (c) effect on nitrate-potentiated twitch. [A,B reproduced by permission from T. Ishizuka, T. Iijima, & M. Endo (1983). *Proceedings of the Japan Academy, 59,* 97–100.]

thetics have an action on the calcium-induced calcium release mechanism similar to that of caffeine (Matsui & Endo, 1986). Furthermore, it was shown that, given the magnitude of enhancement of calcium-induced calcium release in diseased muscle by an anaesthetic concentration of halothane, it is possible to explain the muscle contracture only in diseased muscle under halothane, but not in diseased muscle without halothane or in normal muscle with halothane (Endo et al., 1983). These results conclusively indicate that malignant hyperthermia is a disease of enhanced calcium-induced calcium release.

Ryanodine receptor and the calcium-induced calcium release channel: possible dual modes of operation of the calcium release channel

Recently biochemists have shown that ryanodine, an alkaloid known for a long time to induce contracture in skeletal muscle but to inhibit cardiac contraction, acts on the SR calcium release channel and fixes the channel in an open state (Fleischer et al., 1985). Since the affinity of this

B

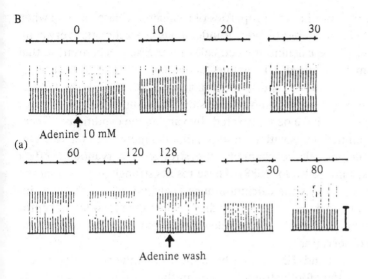

(a)

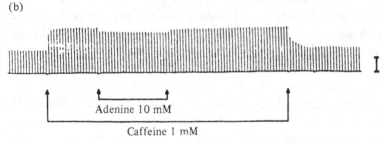

(b)

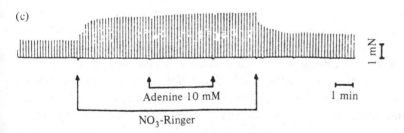

(c)

alkaloid for the channel is very high, it has been utilized to isolate the calcium release channel. The ryanodine receptor protein has successfully been purified and has been found to be larger than 450 kD (Imagawa et al., 1987; Inui, Saito, & Fleischer, 1987; Lai et al., 1988; Saito et al., 1988). Recently, the primary structure of rabbit ryanodine receptor has been determined (Takeshima et al., 1989). The purified ryanodine receptor has the properties of a calcium release channel when incorporated into a lipid bilayer; it indeed has characteristics very similar to those of the calcium-induced calcium release mechanism, in that (1) calcium is required for channel activation and (2) ATP strongly enhances the activation, and so on (Imagawa et al., 1987; Hymel et al., 1988; Lai et al., 1988). Ryanodine acts on the channels only when they are open, and we have demonstrated, by varying the conditions of incubation of skinned fibres with ryanodine, that the more open the calcium-induced calcium release channels, the stronger the ryanodine effect (Oyamada, Iino, & Endo, 1988). These results strongly suggest that the ryanodine receptor is the calcium-induced calcium release channel. Indeed, one of the monoclonal antibodies against the ryanodine receptor inhibits calcium-induced calcium release (M. Iino & H. Takeshima, unpublished observation).

On the other hand, EM studies have shown that the ryanodine receptor looks like the "foot" structures seen in the junction between the T-tubule and the SR, which is thought to be the physiological calcium release channel (Franzini-Armstrong, 1970a). Yet, as described in the preceding sections, the calcium-induced calcium release is not the physiological calcium release mechanism. Thus, it seems likely that one and the same protein molecule, the ryanodine receptor, operates in two entirely independent modes in response to different stimuli – that is, (1) stimulation through the T-tubule depolarisation, and (2) direct stimulation by Ca^{2+} – although the possibility that there are two kinds of calcium release channels that are very similar but slightly different from each other still cannot be excluded. Further studies are urgently needed.

It may be relevant to the present problem to mention the effects of dantrolene sodium (DAN) on calcium release mechanisms here. DAN is well known to inhibit physiological calcium release in skeletal muscle in the micromolar range without affecting action potentials or calcium activation of the contractile mechanism (Ellis & Bryant, 1972; Ellis & Carpenter, 1972; Brocklehurst, 1975; Desmedt & Hainaut, 1977). DAN is also known to be a very effective agent for the therapy and prevention of human and porcine malignant hyperthermia (Harrison, 1975; Kolb, Horne, & Martz, 1982). The latter effect of DAN might suggest that the drug also inhibits the calcium-induced calcium release mechanism since malignant hyperthermia is considered to be due to hyperfunction of calcium-induced calcium release as described in the previous section. However, the results reported on the effects of DAN on calcium-, caffeine-, or halothane-induced calcium release have been controversial:

while some have reported an inhibition (Anderson & Jones, 1976; Austin & Denborough, 1977; Ohnishi, Taylor, & Gronert, 1983), others have reported no effects (Nelson, 1984; Araki et al., 1985). We have found an interesting fact which may resolve the controversy: that DAN strongly inhibits calcium-induced calcium release as well as caffeine- and halothane-induced calcium release at 37°C, but does not inhibit them at all at room temperature (Fig. 5.6; Ohta & Endo, 1986; Kobayashi & Endo, 1988). As shown in Fig. 5.7, the dose–inhibition relation of DAN on caffeine-induced calcium release is very similar to that on calcium release induced by T-tubule depolarisation at 37°C, which is consistent with the idea that the sites of action of DAN in these two effects are indeed one and the same. This view is supported by the fact that all the derivatives of DAN so far examined showed similar dose–inhibition relationships on both calcium release mechanisms as in the case of DAN, although their potencies are widely different from one derivative to another. In other words, among DAN derivatives we have not so far succeeded in separating the inhibitory action on caffeine-induced calcium release from that on physiological calcium release (T. Yamamoto, T. Kobayashi, & M. Endo, unpublished results). On the other hand, at

Fig. 5.6. Effect of 50 μM dantrolene (Dant) on the rate of calcium release induced by pCa 5.5 with or without caffeine. Caffeine was present only immediately before and during calcium-induced calcium release. Mean $\pm$ S.E.M. ($n = 4$). *$p < .01$. [Reproduced by permission from T. Ohta & M. Endo (1986). *Proceedings of the Japan Academy, 62,* 329–32.]

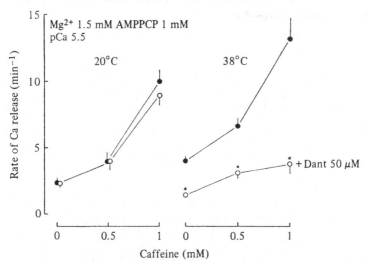

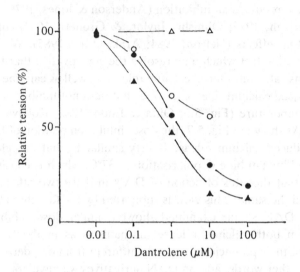

Fig. 5.7. Dose–inhibition relationships of dantrolene on caffeine- and K-induced contractures at different temperatures. Circles: K contractures; triangles: caffeine contractures. Open symbols: at 20°C; filled symbols: at 37°C. [Reproduced by permission from T. Kobayashi & M. Endo (1988). *Proceedings of the Japan Academy, 64*, 76–9.]

room temperature DAN still inhibits physiological calcium release with similar potency, but it no longer inhibits calcium- and caffeine-induced calcium release as already mentioned (Fig. 5.7). These facts are very interesting, especially in view of the proposed dual mode of operation of the ryanodine receptor protein.

6

Hypodynamic tension changes in the frog heart

ROLF NIEDERGERKE AND SALLY PAGE

Introduction (by R. Niedergerke)

I joined A. F. Huxley's laboratory in the autumn of 1952, coming from
A. von Muralt's Institute in Berne, where I had done some work on
single myelinated nerve fibres. Contact with Cambridge had been established through Robert Stämpfli, from whom I had learnt the nerve fibre
dissection, and who hoped that I could also manage the muscle fibres
required for A.F.H.'s next research project. Banking on this hope, we
had made plans for a collaboration. Another reason I was not entirely
unprepared for this venture goes back to the time of my previous position as a demonstrator in physiology at Göttingen, where I had occasion
(in preparing a seminar) to search the literature for information on the
physiological function of the striations in skeletal muscle. The yield had
been disappointingly meagre and confusing, the latter because the
changes in the muscle's band pattern during contraction were both controversial and complex. That I was destined to help clarify this matter in
the near future I could not have suspected, nor that I should meet
someone of A.F.H.'s calibre. He had delved into the physics of the
image formation due to transmitted light of the unstained muscle fibre,
and had come to the conclusion that results obtained using the ordinary
light microscope on unstained vertebrate muscle were unreliable, probably even misleading (for an account of the historical background, see
A. F. Huxley, 1977, 1980a). He had designed his own interference micro-

Support of this work by a Leverhulme Emeritus Fellowship Award to R.N. is
gratefully acknowledged.

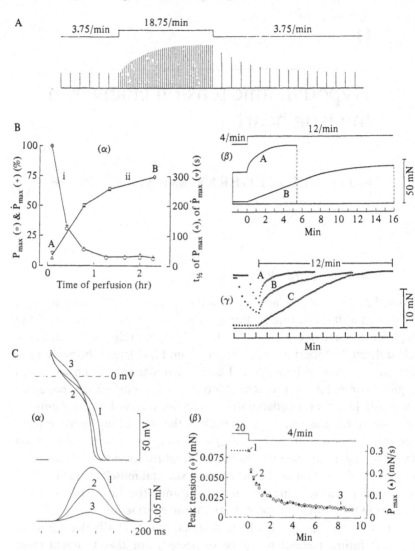

Fig. 6.1. Staircase and development of the hypodynamic state of the frog heart. (A) Twitch response to enhanced and subsequent return to original stimulus rate as indicated; ventricle strip, *Rana pipiens*. (B) (α) *Curve i:* twitch changes after commencement of perfusion of recently dissected ventricle with 1 mM [Ca] Ringer's, regular stimulation at 4 min⁻¹; peak twitch tension, P (O), and maximum rate of tension rise, $\dot{P}_{max}$ (+), both expressed as a percentage of initial value (left ordinate). *Curve ii:* half-times of ascending staircases during periods of enhanced stimulus rate, at 12 min⁻¹; peak twitch tension (△), maximum rate of tension rise (▽) (right ordinate). (β) Profiles of ascending staircases (twitch tension), same experiment as in (α), A at beginning of perfusion, B more than 2 h later. (γ) Ascending staircases,

scope, building much of the prototype himself, and on my arrival was (almost) ready to assemble version No. 2 with which to carry out the planned experiments. Progress was swift: after construction of further apparatus and barely a year of experimentation, our evidence for the "sliding filament mechanism" of contraction was submitted for publication, together with that obtained by H. E. Huxley and Jean Hanson, who had worked simultaneously and independently on the same project, albeit with different techniques (A. F. Huxley & Niedergerke, 1954; H. E. Huxley & Hanson, 1954). Thereafter, two years of work (with other tasks alongside) brought forth A.F.H.'s first cross-bridge model of contraction (A. F. Huxley, 1957a), encapsulating most of what was known about the mechanical and energetic properties of skeletal muscle. In the meantime, I had turned to an examination of the effects of calcium, first on skeletal muscle, while still in Cambridge (Niedergerke, 1955), and later in London on the frog heart (Niedergerke, 1956). (The move to a different muscle, suggested by Bernard Katz, was to be crucial for my future work.) Oddly enough, the first project I undertook after Cambridge dealt with a problem not yet solved, but also not forgotten, as indicated in the discussion to follow.

The hypodynamic state

In his influential paper, A. J. Clark (1913) wrote: "A frog heart, when isolated and perfused for a few hours with Ringer's fluid, passes into an enfeebled state in which the force of contraction is impaired (the hypodynamic state)". Curve i of Fig. 6.1B (α) shows this gradual decline of both peak tension, P, and a maximum rate of tension rise, $\dot{P}_{max}$, of twitches during the first 2 h of perfusion. (N.B. The low stimulus rate used, of 4 min^{-1}, accentuates the phenomenon.) Since the initial strength of the heartbeat could readily be restored by means of excess calcium,

Caption to Fig. 6.1. *(cont.)* stimulus rate increased from 4 to 12 min^{-1} at 1 min (A), 2 min (B), and 20 min (C) after start of preceding descending staircase, 2 mM $[Ca]_o$ (different ventricles used in γ and α, both *Rana pipiens*). (C)(α) Tracings of cellular action potentials and tension records during descending staircase in (β) (stimulus rate = 4 min^{-1}, reduced from 20 min^{-1} at zero time); (○) left ordinate, peak tension; (▲) right ordinate, maximum rate of tension rise; ventricular trabecula, *Rana temporaria*. [A reproduced by permission from semischematic illustration of J. Koch-Weser & J. R. Blinks. (1963). The influence of the interval between beats on myocardial contractility, *Pharmacological Reviews, 15,* 601–52, copyright by The Williams & Wilkins Company. B adapted from Chapman & Niedergerke, 1970b.]

reduced sodium, or reduced potassium in the external medium, effects previously described by Ringer (1883), this contractile failure seemed unlikely to be due to muscle fatigue. Instead, Clark surmised that a substance was washed out from heart cell surfaces, and, as a result, "the heart loses the power of combining with calcium". At about the same time, Boehm (1914) and Lieb and Loewi (1918) reported the loss of calcium ions from frog hearts perfused with artificial saline.

A phenomenon much dependent on the hypodynamic condition of preparations is the staircase (Fig. 6.1A), the response to the increase in heart rate discovered by Bowditch (1871). As the transformation of curve A to B of Fig. 6.1B(β) shows, the buildup of the staircase becomes slow as the heart passes into the hypodynamic state. Various ways of restoring the original appearance of the staircase are known, for example: increasing external calcium, or reducing sodium or potassium concentrations – that is, the changes mentioned above. Also, the staircase depends critically on the preceding history of the preparation: its buildup is rapid when it is evoked early during the decline of a preceding staircase [curve A, Fig. 6.1B(γ)] but becomes slow when evoked later on (curves B & C, same panel).

For completeness, Fig. 6.1C(α) includes tracings of cellular action potentials recorded during a descending staircase [shown in (β)]: overshoot and plateau potentials, previously depressed, recover while action potential durations shorten – changes related presumably to the declining levels of cellular [Ca] (perhaps also [Na]). (For analysis of similar, though more complex changes in the mammalian heart, see Hilgemann & Noble, 1987.)

Figure 6.2A shows the calcium effect, the parallel increase of twitch tension and maximum rate of tension rise via an initial rapid and subsequent slow phase at high (4 mM) $[Ca]_o$ and the decline along a composite time course of a similar kind on return to the original low (2 mM) $[Ca]_o$. Similar to the staircase, the buildup of this reponse is rapid in the fresh heart (not shown), but phases of slow tension change appear during development of the hypodynamic condition: in a state of deep hypodynamia, twitch buildup in high $[Ca]_o$ is predominantly slow (right-hand panel of Fig. 6.2A). Another property the calcium response shares with the staircase is its dependence on the foregoing treatment of the preparation: when evoked shortly after a preceding response, its buildup is rapid but becomes slower when evoked later on (Fig. 6.2B, runs plotted as open circles).

Thus, development of the hypodynamic condition is accompanied by

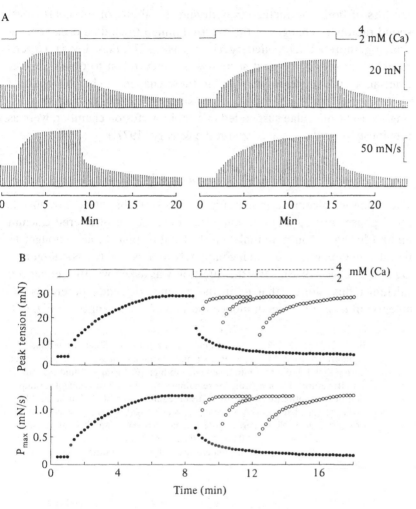

Fig. 6.2. Properties of calcium response of frog heart ventricle. (A) Twitch tension (*upper panel*) and maximum rate of tension rise (*lower panel*) due to $[Ca]_o$ changes indicated; result on the right obtained about 30 min after that on the left; stimulus rate at 4 min^{-1}; *Rana pipiens*. (B) (●) Changes of twitch tension and maximum rate of tension rise after $[Ca]_o$ steps from 2 to 4 mM and back to 2 mM; (○) data from three other runs, 4 mM $[Ca]_o$ reapplied at times shown by dashed lines above graph after the preceding return to 2 mM [Ca] Ringer's; stimulus rate at 5 min^{-1}; ventricle, *Rana pipiens*. (Adapted from Chapman & Niedergerke, 1970a.)

the loss of both contractile strength and the ability of the heart to respond rapidly to changes of the external milieu (including drug application, e.g., adrenalin, cf. Niedergerke & Page, 1977, Figs. 1 & 4). Experiments discussed in the next section were undertaken to obtain further clues about the processes involved in these changes. (The conventional methods used for the recording of tension and membrane potential, in single atrial trabeculae suspended in a rapid perfusion chamber, were as previously described; see Niedergerke & Page, 1977.)

Modification of slow twitch changes by cellular calcium uptake

The biphasic calcium response (Fig. 6.2) is probably of dual origin, its rapid phases arising as the immediate consequence of altered calcium influx (during action potentials), with the slow phases, also thought to be calcium dependent, as yet incompletely understood (cf. Niedergerke, Ogden, & Page, 1976). First the question was asked whether enhanced calcium influx alone – that is, in the absence of action potentials – is able to influence or, indeed, induce the slow phasic changes. In Fig. 6.3,

Fig. 6.3. High $[Ca]_o$ exposure of quiescent preparation. Single trabecula, equilibrated in 2 mM [Ca] Ringer's (regular stimulation at 10 min^{-1}), exposed during first 4.7 min of 5-min quiescence to high [Ca] Ringer's fluid (as indicated); return to 2 mM $[Ca]_o$ for remaining 0.3 min and subsequent resumption of stimulation; half-time of diffusion equilibration of [Ca] at heart cell surfaces, 1.8 s [from $t_{1/2}$ of twitch changes after $[Ca]_o$ steps at regular stimulation (cf. Page & Niedergerke, 1972)]; hence, the calculated [Ca] at cell surface after 20 mM $[Ca]_o$, 2.02 mM at time of renewed stimulation; atrial trabecula, *Rana temporaria;* temperature in this and all other experiments, 20–22°C.

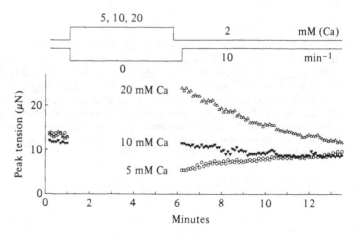

an atrial trabecula was exposed during a 5-min period of quiescence to Ringer's fluid of enhanced [Ca] (5, 10, or 20 mM), followed by recovery in normal Ringer's fluid in which the effects of this treatment were assessed from the twitch changes obtained on renewed stimulation. As is seen, twitches were markedly increased after 20 mM [Ca], hardly if at all after 10 mM, whereas 5mM [Ca] failed to prevent the decline normally occurring during quiescence (not shown). Any subsequent twitch change, decline or buildup, was slow ($t_{1/2}$ of 100–200 s).

In the experiment shown in Fig. 6.4, 10 mM [Ca] Ringer's fluid (re-

Fig. 6.4. Modification of slow twitch decline by $[Ca]_o$ pulse. (A) Peak tension (*top graph*) and maximum rate of tension rise, $\dot{P}_{max}$ (*bottom graph*) of twitches during ascending followed by descending staircase (first part, semischematic), stimulus rates as shown, 1 mM $[Ca]_o$; during descending staircase, exposure to 10 mM $[Ca]_o$ for first 8 s of 36-s silent interval. (B) Semilogarithmic plot of data of descending staircase of (A), as differences of P and $\dot{P}_{max}$ with respect to final steady levels, P_{final} and $\dot{P}_{max,final}$, (abbreviated to ΔP and $\Delta \dot{P}_{max}$), in units of differences at $t = 0$; straight lines fitted by eye; ($[Ca]_o$ level at heart cell surfaces at end of quiescence, 1.1 mM, computed as described in legend of Fig. 6.3. N.B. Preparation highly sensitive to $[Ca]_o$). Atrial trabecula, *Rana temporaria*.

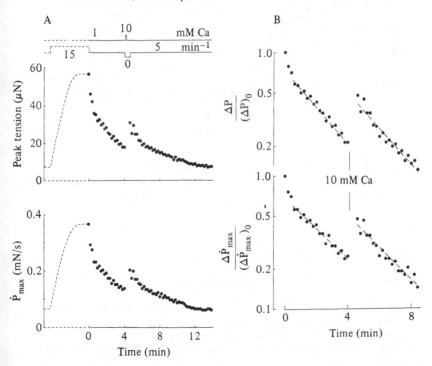

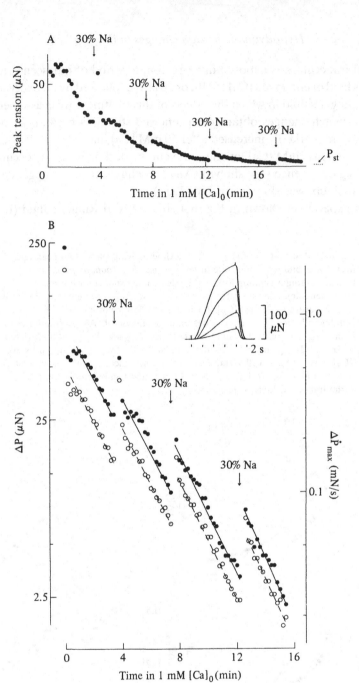

Fig. 6.5. Slow twitch decline modified by low $[Na]_o$ steps. (A) At $t = 0$, $[Ca]_o$ reduced from 3 to 1 mM (steady stimulus rate of 5 min^{-1}); at arrows, low, 30% $[Na]_o$ exposure during first 8 s of 24-s silent intervals (NaCl replaced by sucrose; [Na] at heart cell surfaces at end of silent interval, 98.8% of normal,

placing 1 mM [Ca] Ringer's) was briefly applied within a silent interval interposed during a period of slow twitch decline, itself the response to a reduced stimulus rate (details in legend). Here the twitch increase caused by high $[Ca]_o$ was succeeded by a decline at approximately the same rate as that of the descending staircase preceding the high [Ca] step (see data of P and $\dot{P}_{max}$ in the semilogarithmic plot of Fig. 6.4B).

In a different but related experiment (Fig. 6.5), a low [Na] medium applied to a trabecula at various stages of slow twitch decline served to induce brief increases of calcium influx through the Na–Ca exchanger of the cell membrane. As is shown (Fig. 6.5B), twitch tension was increased after each low [Na] step before continuing to decline at (approximately) the former rate. It is shown further (inset of Fig. 6.5B) that the strength of contractures evoked by the low [Na] medium declined, too, alongside that of the twitches.

Clearly, results of Figs. 6.3 to 6.5 suggest that the processes responsible for the slow twitch changes are, indeed, calcium dependent and do not require additional ion fluxes of the action potential. On the other hand, the similarity of the decline of twitch and contracture tension in Fig. 6.5 raises the question whether parallel changes of this kind also occur during tension buildup. In one of the experiments undertaken to test this (Fig. 6.6) a trabecula was subjected to alternate 5-s periods of "systoles" in 25% [Na] Ringer's and 10-s periods of "diastoles" in 100% [Na] Ringer's, respectively. The ascending peaks of contractures so evoked are plotted in curve i of Fig. 6.6A, and tracings of five such responses (encircled symbols) are superimposed in Fig. 6.6B. When, after contracture buildup to 63% maximum tension, electrical stimulation was resumed, twitch tension was first increased in the normal (1 mM [Ca]) Ringer's fluid before slowly subsiding to its original low level (with a $t_{1/2}$ of 2.5–3.0 min; see curves i in the semilogarithmic plot of data of P and $\dot{P}_{max}$ in Fig. 6.6C). In another run of this experiment, a twitch decline in 1 mM [Ca] medium (curves ii of Fig. 6.6C) succeeded the buildup, also of twitches, in a 3 mM [Ca] medium (curve ii, Fig. 6.6A). The similarity of the two pairs of declining curves under comparison

Caption to Fig. 6.5. *(cont.)* obtained as for [Ca], in legend Fig. 6.3, using, in addition, diffusion coefficients of 1.35 and 0.8 × 10^{-5} cm$^2 \cdot$ s^{-1} for Na$^+$ and Ca^{2+}, respectively); P_{st}, steady twitch tension at 1 mM $[Ca]_o$. (B) Semilogarithmic plot of twitch decline in (A) of peak tension (●, left ordinate) and maximum rate of tension rise (○, right ordinate), with respect to steady final levels. (Symbols at zero time from last twitch in 3 mM $[Ca]_o$.) *Inset:* superimposed tracings of first four contractures, of decreasing strength, in the low [Na] medium. Atrial trabecula, *Rana temporaria*.

(Fig. 6.6C) suggests that the twitch and contracture buildups preceding them were due, at least in part, to the same process(es).

Two more detailed points are these: (1) Results were often more complex than just discussed. For example, in Fig. 6.6D, after a series of contractures similar to that of curve i in Fig. 6.6A, a subsequent twitch decline was preceded by a transient buildup, perhaps indicative of rapid recovery from some kind of contracture fatigue. (2) Contracture buildup was obtained also with high [Ca] in place of low [Na] media – for example, in a series of 15-s periods of exposure to 30 mM $[Ca]_o$, separated by 15-s periods in 1 mM [Ca] medium.

The role of cellular calcium "sinks"

A mechanism previously suggested for the slow twitch changes relates to cellular binding sites for calcium other than those specifically concerned with contractile activation (Robertson, Johnson, & Potter, 1981; Campbell et al., 1988a; see also Niedergerke, Ogden, & Page, 1976). Calcium "sinks" of this kind are made up chiefly of the unspecific (i.e., non-contraction-inducing) sites of troponin, as well as sites on calmodulin, cell membrane, and membranes of sarcoplasmic reticulum and other cellular organelles. Because of the high affinity for calcium of most of these sites, a fraction of calcium entering cells during the action potential is expected to combine with these rather than the specific sites. However, when diastolic $[Ca^{2+}]_i$ is high, as after exposure of preparations to high $[Ca]_o$ or high stimulus rates, many of these unspecific sites would be occupied and therefore unable to compete with the specific ones for calcium, so allowing more effective tension activation. (For variation of diastolic $[Ca^{2+}]_i$ with heart rate and $[Ca]_o$, see Lado, Sheu, & Fozzard, 1982; Sheu & Fozzard, 1982; Harding et al., 1989; Frampton, Orchard, & Boyett, 1990.) Evidence for a mechanism of this kind was obtained with a trabecula subjected to two different conditioning treatments (Fig. 6.7A): in the one, serving to establish a state of accentuated hypodynamia (at low $[Ca]_o$ and stimulus rate, see legend), tests (run i) with two consecutive low $[Na]_o$ steps showed that the second of the two contractures so evoked started with a 3-s shorter latency (varying from 2.5 to 3.5 s in 4 runs) than the first (see displaced first response, dashed line, superimposed upon the second). By contrast, after the alternative conditioning, made to alleviate hypodynamia (with high $[Ca]_o$ and stimulus rate), this latency difference was absent, as the two runs ii and iii of

Fig. 6.7A illustrate (details in legend). A likely explanation is that calcium taken up by cells during the first response of run i had replenished a calcium sink that had already been filled during conditioning in runs ii and iii. In agreement with this idea, the latency became shorter (1.0–1.5 s

Fig. 6.6. Contracture staircase and its after-effects. (A) *Curve i:* contracture peaks during cycles of successive exposures, 5.0 s (±0.5 s) to 25% [Na] and 10 s to 100% [Na] Ringer's, 1mM [Ca] in both. *Curve ii:* twitch peaks from another run after increasing [Ca]$_o$ from 1 to 3 mM; regular stimulation at 5 min^{-1}. P_m (R.H. ordinate), maximum tension of trabecula. (B) Superimposed tracings of contractures from (A) marked by encircled symbols. (C) Semilogarithmic plot of twitch decline in 1 mM [Ca]$_o$ – that is, differences, ΔP (filled symbols) and $\Delta \dot{P}_{max}$ (open symbols) as in legend of Fig. 6.4, *run i,* after contracture buildup, and *run ii,* after twitch buildup of (A). (D) Twitch decline as in (C) i, after contracture buildup like that of run i in (A), but from different trabecula (alternate twitches plotted). Units of ΔP, μN, of $\Delta \dot{P}_{max}$, mN·s^{-1} in both (C) and (D). Atrial trabeculae, *Rana temporaria;* stimulation at 5 min^{-1} in both.

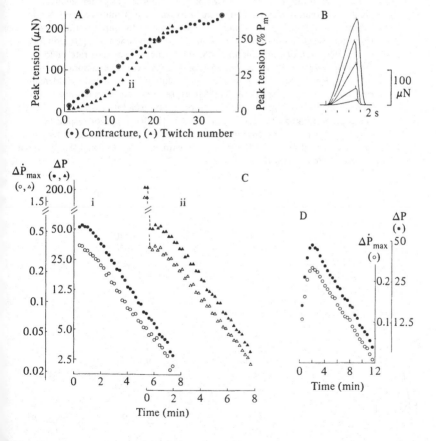

instead of 3 s) after a conditioning treatment intermediate, as regards cellular calcium loading, between that of runs i and ii (details in legend, but results not illustrated).

Does the filling and emptying of a cellular calcium sink provide the

Fig. 6.7. Tests for effects of cellular calcium sink. (A) *Run i:* conditioning of trabecula, 8-min exposure to 0.25 mM [Ca] Ringer's, stimulation at 5 min^{-1}; 12 s before start of trace, $[Ca]_o$ increased to 1.5 mM and maintained for subsequent test; first (7 s) exposure to 25% [Na] medium separated from second (in same medium) by 12-s recovery period in 100% $[Na]_o$; dashed trace, record of first response shifted to start of second. *Runs ii and iii:* conditioning in 2 mM [Ca] Ringer's, stimulation at 15 min^{-1}; first exposure to 50% $[Na]_o$ for 3 s in ii, 6 s in iii, second exposure to same medium after 12-s recovery in 100% $[Na]_o$, 2 mM $[Ca]_o$ throughout; dashed traces constructed as for run i. Horizontal dotted lines: final level of second response. [N.B. Low (50%) [Na] medium of runs ii and iii and durations (3 s and 6 s) of first exposure were chosen so that (a) tension rise of second responses in ii and iii was similar to that in i, and (b) size of first response was either smaller, ii, or greater, iii, than that in i.] Conditioning for the fourth run (see text), 8-min exposure to 0.5 mM [Ca] medium, stimulus rate of 15 min^{-1}. Atrial trabecula, *Rana temporaria*. (B) Another test. Two conditioning treatments, (1) equilibration of trabecula in 2 mM [Ca] Ringer's, stimulus rate of 15 min^{-1} (for state of alleviated hypodynamia, runs a,c); (2) equilibration in 1 mM [Ca] Ringer, stimulus rate of 5 min^{-1} (state of accentuated hypodynamia, runs b,d); test responses all in 2 mM $[Ca]_o$ (established for conditioning (2) 12 s before start of trace). Test media for responses a [after conditioning (1)] and d (conditioning (2)], 100% [Na], 50 mM [K] Ringer's, for responses b [conditioning (2)] and c [conditioning (1)], 100% [choline] Ringer's, followed (at arrows) by 100% [Na] Ringer's, both with 50 mM additional [KCl]. Atrial trabecula, *Rana temporaria*.

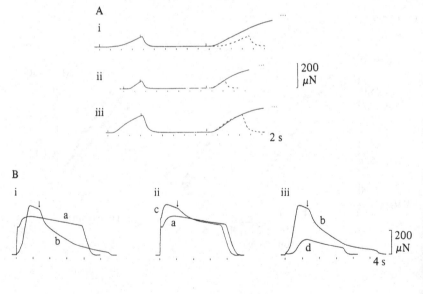

full explanation for the slow twitch changes? To examine this question (Fig. 6.7B), a trabecula was again conditioned to be in a state of either alleviated (1) or accentuated (2) hypodynamia [using (1) enhanced or (2) low $[Ca]_o$ and stimulus rates, respectively; see legend]. After conditioning (1) (record a), exposure of the trabecular to 100% [Na] medium caused a rapid tension rise including a twitch, to a shallow summit followed by a slow "spontaneous" decline. (N.B. All test fluids contained excess 50 mM [K].) After conditioning (2) (accentuated hypodynamia), the tension rise in nominally zero [Na] medium (record b) was at first slow, presumably during replenishment of a cellular calcium sink, and then accelerated to a peak higher than that of record a, before giving way to a slow decline. This was followed, in 100% [Na] medium (applied at time marked by arrow), by a further, initially steep, decline to tension levels much below those of response a. Since calcium occupation of troponin sites, specific and unspecific, should have been greater at the peak of response b than a, yet failed to prevent the large tension drop of b, an additional process besides filling of a (readily available) cellular calcium sink seems to have caused the better maintained tension of response a. Panels ii and iii of Fig. 6.7B illustrate controls made to check whether some kind of contracture fatigue could have produced some of the features described, especially those of response b. A comparison was made, after both conditioning treatments, of the effects of the two different test media used: 100% [Na] Ringer's maintained throughout runs a and d, and nominally zero [Na] medium present for the initial 5 s, followed by 100% [Na] Ringer's for the remainder of the two runs b and c. As is seen in both ii and iii, contracture peaks at low, nominally zero $[Na]_o$ (c & b) were succeeded by a tension fall in subsequent 100% $[Na]_o$ to levels close to those also attained during maintained exposure to this medium (records a & d). Thus, effects of contracture fatigue, if occurring at all, cannot have been important.

The role of the sarcoplasmic reticulum

It is widely held that twitch staircases are related to calcium transfer into (or out of) the sarcoplasmic reticulum (SR), a possibility that has been examined in the following experiments. (For a preliminary note, see Niedergerke & Page, 1979.) These were carried out in two series of recycled runs; in the controls (curve i, Fig. 6.8) a slow twitch buildup was initiated in a medium of enhanced $[Ca]_o$, which caused the usual biphasic response. In the main run (curve ii), this calcium effect was combined with that of caffeine, at the same time to facilitate SR calcium discharge and depletion of the SR store. Thus, the initial large upward swing of curve ii, not seen in curve i, is readily understood to be due to enhanced SR calcium discharge; the subsequent decline, to SR store depletion (see Niedergerke & Page, 1981); and the then ensuing slow tension increase, to (at least in part) the same process also causing curve i to rise.

Results of the second part of the experiment (at $t > 5.5$ min), after reapplication of the original 1.5 mM [Ca] medium and the withdrawal of caffeine (in the case of run ii), show that twitch decline in the two runs occurred with a similar time course, via the familiar two consecutive

Fig. 6.8. Are slow twitch changes related to SR function? *Curve i* (control): twitch buildup due to $[Ca]_o$ change from 1.5 to 3.5 mM. *Curve ii:* same $[Ca]_o$ change as in i but with additional 3 mM caffeine; both i and ii are succeeded on the right by recovery in 1.5 mM [Ca] Ringer's; length of vertical bar at t = 7.6 min of curve ii, 2 × S.D. of data from 5 runs. (For clarity, data of alternate twitches in 3.5 mM $[Ca]_o$ and of every fourth twitch in 1.5 mM $[Ca]_o$ are plotted, except for those immediately after change in $[Ca]_o$.) *Curve iii:* procedure as for curve ii but with recovery periods of 1.5 mM $[Ca]_o$ of different durations (indicated by dotted lines above graph), ending with single test twitch in 3 mM caffeine, 0.1 mM [Ca] medium after 12-s (quiescent) period of exposure to this medium; circled data points, twitches superimposed in inset A. Inset B, twitch recoveries from different experiment but with procedure closely similar to that of main graph. Curve iv, control. Curve v, after 5 mM caffeine. Atrial trabeculae, *Rana temporaria* in both; regular stimulation at 10 min^{-1} in main graph, 15 min^{-1} in inset B.

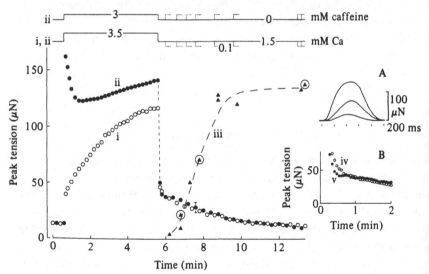

phases, rapid and slow. Data from additional runs, curve iii, were used to monitor the calcium content of the SR store, by means of a "test" twitch procedure: run ii was recycled up to time 5.5 min and followed by recovery periods in 1.5 mM [Ca] Ringer's of varying durations (from 0.33 to 5 min), each ending with a single twitch in caffeine medium (applied for a 12-s period at reduced, 0.1 mM $[Ca]_o$, the latter to keep the component of twitch activation due to calcium influx small; further details in legend). The buildup of these twitches (superimposed tracings of three in inset A) as indicative of the recovery of the previously reduced SR calcium content (Niedergerke & Page, 1981). Its concurrence with the twitch decline in normal Ringer's fluid suggests that SR activity and the process(es) causing the slow tension decline are separate events.

The twitch recoveries shown in curves iv and v (inset B), similar to those of curves i and ii of the main graph but from a different experiment, illustrate an additional feature: the initially more rapid decline of curve v after caffeine exposure than that of curve iv, the control, suggests that the caffeine effect was succeeded by a transient twitch decline. Transients of this kind were observed in conditions of constant external [Ca] as after-effects of α-adrenalin and ATP action (Niedergerke & Page, 1989) and attributed to an effect of preceding SR store depletion, an explanation applicable to this result also.]

Discussion

Before discussing the results in greater detail, it is worth noting that slow twitch changes occur, even in preparations that do not succumb to the hypodynamic condition (a small minority) when they are exposed to sufficiently low [Ca] media (e.g., 0.5 to 1 mM instead of the more usual 1.5 to 2 mM). Rapid twitch responses characteristic of the pre-hypodynamic condition are then restored, simply by re-equilibrating the preparations in normal Ringer's fluid.

With regard to the origin of the slow changes, the outcome of Figs. 6.7 and 6.8 suggests that varying calcium binding to unspecific sites of the cytoplasm and a varying calcium content of the SR provide only partial explanations, and that additional factors seem to be at play. These may be discussed in conjunction with Fig. 6.9, which illustrates another aspect of the slow twitch changes, the concurrent sensitivity change of preparations to external calcium (see also Chapman & Niedergerke, 1970a). The procedure in this experiment (Fig. 6.9A) had been to apply small (0.25 mM), consecutive positive and negative $[Ca]_o$ steps to a

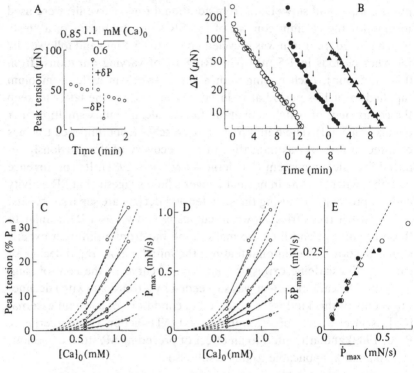

Fig. 6.9. Sensitivity change to $[Ca]_o$ during slow twitch decline. (A) Illustration of procedure: during twitch decline in 0.85 mM [Ca] Ringer's, application of +0.25 mM and −0.25 mM $[Ca]_o$ steps, each for one beat interval (12 s); $+\delta P$ and $-\delta P$, increment and decrement of twitch peak with respect to average P before and after each step. (B) Semilogarithmic plots of ΔP (definition in legend of Fig. 6.4) from three runs, all in 0.85 mM $[Ca]_o$, after twitch buildup in (○) 2 mM [Ca] Ringer's, (●) 1 mM [Ca]–70.6% [Na] Ringer's (NaCl replaced by sucrose), and (▲) 0.85 mM [Ca]–1 mM [K] Ringer's. Arrows: times of $[Ca]_o$ steps (last steps, unmarked, at steady twitch tension in all runs); lines drawn by eye (data of only alternate twitches plotted). (C,D) Sets of twitch data connected by continuous lines for P in (C) and $\dot{P}_{max}$ in (D) from $[Ca]_o$ steps after buildup in 2 mM $[Ca]_o$; dashed lines were computed by means of eq. (6.1): $P = 7.5[(X \cdot [Ca]_o) - 0.4]^2$, with X of 1.0, 1.31, 1.53, 1.80, 2.15, and 2.51, respectively, in (C); and eq.(6.2): $\dot{P}_{max} = 2.45Y([Ca]_o - 0.15)^2$, with Y of 1.0, 2.1, 3.0, 4.4, 6.7, and 9.0, respectively, in (D); (E) plot of $(+\delta\dot{P}_{max} + |-\delta\dot{P}_{max}|)/2$ against $\dot{P}_{max}$, average $\dot{P}_{max}$ in 0.85 mM $[Ca]_o$ for each set; conditions and symbols as in (B); dotted line obtained with $\delta\dot{P}_{max} = 0.71\ \dot{P}_{max}$. Stimulus rate of 5 min⁻¹; atrial trabeculae, *Rana temporaria*.

trabecula at various stages of twitch decline in a low 0.85 mM [Ca] medium (after previous equilibration in 2 mM [Ca] Ringer's). Sets of values of P and $\dot{P}_{max}$ at 0.6, 0.85, and 1.1 mM $[Ca]_o$ so obtained are shown in panels C and D, and averages of pairs of increments and decrements $[(+\delta\dot{P}_{max} + | -\delta\dot{P}_{max}|)/2]$ in panel E (open circles), the latter plotted against levels of $\dot{P}_{max}$ at 0.85 mM $[Ca]_o$, above and below which they were determined. The change in slope of the curves so generated (at $[Ca]_o$ of 0.85 mM; panels C and D) was about 10-fold, and up to 15-fold in other experiments of this kind, six in all. It reflects, at any rate in part, the sensitivity loss of the preparation to $[Ca]_o$. Similar changes occurred during twitch decline after a preceding buildup in media of reduced [Na] or [K] (the filled circles and triangles of panel E), a similarity extending to the time course of these twitch declines (panel B, the semilogarithmic plots of tension changes after high $[Ca]_o$, low $[Na]_o$, and low $[K]_o$ treatment, respectively). Apparently, thus, the processes under discussion are similar in three conditions.

Two alternative mechanisms, both slow, are invoked (dashed curves of panels C and D) to explain these results: the one modifies the calcium quantity transferred during the action potential to myofibrils [parameter X, eq. (6.1)]; the other [parameter Y, eq. (6.2)] alters the sensitivity of myofibrils to calcium so transferred. N.B. Only one of several possible schemes of this type is being considered:

$$P \propto (X[Ca]_o - b)^2 \qquad (6.1)$$

$$P \propto Y([Ca]_o - b)^2 \qquad (6.2)$$

Assumptions made in writing eqs. (6.1) and (6.2) are that (1) the relationship between tension and cellular calcium can be approximated by a parabola in the tension range used (as was the case for the two curves of Niedergerke & Page, 1989, fig. 4C), (2) calcium transfer into cells during action potentials is proportional to $[Ca]_o$; (3) contractile activation occurs when calcium exceeds a "threshold" quantity (proportional to b) which, for simplicity, is taken to be invariant; and (4) no allowance is made for the effects of calcium sinks.

Although the analysis is clearly tentative, nevertheless the satisfactory fit of the two sets of curves to the data points in panels C and D suggests at least the feasibility of alternative interpretations. One of these concerns the modification of calcium fluxes which pass through the Na–Ca exchanger of the cell membrane during the action potential. Because of uncertainty as regards the size of these fluxes and their contribution to

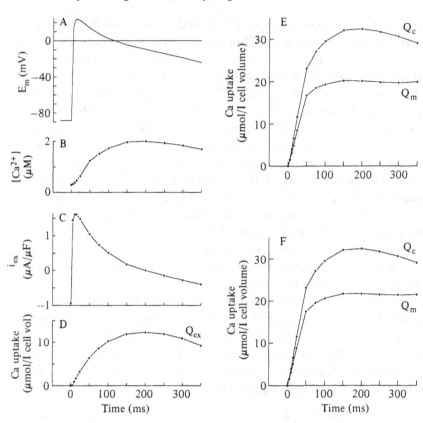

Fig. 6.10. Computed calcium uptake by atrial cells during initial 350 ms of an action potential. (A) Tracing of action potential, E_m, recorded in 2 mM [Ca] medium at a stimulus rate of 10 min^{-1}. (B) Curve of $[Ca^{2+}]_i$, computed as (aequorin signal)$^{0.4}$ from a record, also at 2 mM $[Ca]_o$, of fig. 4 of Allen and Blinks (1978), starting from an assumed diastolic $[Ca^{2+}]_i$ of 3×10^{-7} M, and rising to a peak of 2×10^{-6} M $[Ca^{2+}]_i$ (cf. Allen & Blinks, 1978). (C) Current, i_{ex}, through Na–Ca exchanger obtained from eq.(6.3) (see below), with E_m from (A), $[Ca^{2+}]_i$ from (B), and with 117 mM $[Na]_o$, 2 mM $[Ca]_o$; for $[Na]_i$ and other parameters, see below. (D) Integral of curve (C) with respect to time, with a surface/volume ratio of 1 μm^{-1} (Page & Niedergerke, 1972) and a specific membrane capacitance of 1 $\mu F/cm^2$ to give Q_{ex}, calcium uptake through exchanger (per unit of cell volume). (E) Q_c, calcium uptake from both calcium inward current and exchanger influx, computed from curve (B) using cellular calcium binding sites summarised by continuous curves of fig. 4B of Niedergerke and Page (1989), and Q_m, calcium uptake due to calcium inward current, from the difference between Q_c and Q_{ex}. (N.B. Steady Q_m, ~20 μmol · l^{-1}, is reasonably close to ~10 μmol · l^{-1} of Campbell et al., 1988a, especially when uncertainties of both their determination and the present approach are taken into account.) (F) Q_c, as in (E), but Q_m from Q_{ex}, the integral $\int_0^t i_{ex}dt$,

the total calcium movements during the action potential, especially in physiological conditions, the attempt was made to assess these factors through computation (Fig. 6.10), using data from several sources. First the combined calcium uptake, Q_c, through exchanger and calcium channel (panel E) was obtained from an aequorin light signal (recorded by Allen & Blinks, 1978, converted into $[Ca^{2+}]_i$, panel B), in conjunction with plausible data on calcium binding sites inside the cells (Niedergerke & Page, 1989, fig. 4B). Next the exchanger current (panel C) was

Caption to Fig. 6.10. *(cont.)* with i_{ex} from eq. (6.4), instead of (6.3). Eq. (6.3) [adapted from Mullins (1977, eq. 23), with 3 Na:1 Ca stoichiometry]:

$$i_{ex} = \frac{k'([Na]_i^3[Ca]_o\exp(\gamma E_m F/RT) - [Na]_o^3[Ca]_i\exp\{-(1-\gamma)E_m F/RT\})}{\{(\alpha_i+\beta[Na]_i^3[Ca]_o)(\beta[Na]_o^3[Ca]_i\exp(-(1-\gamma)E_m F/RT)+\kappa) + (\alpha_o+\beta[Na]_o^3[Ca]_i)(\beta[Na]_i^3[Ca]_o\exp(\gamma E_m F/RT) + \kappa)\}} \quad (6.3)$$

assuming that (1) translocation rate constants of unoccupied exchanger ($=x \cdot \kappa$) are identical for in- and outward movement through cell membrane; (2) rate constants of occupied exchanger for in- and outward calcium movement are $x \cdot \exp(\gamma E_m F/RT)$ and $x \cdot \exp(-(1 - \gamma)E_m F/RT)$, respectively (DiFrancesco & Noble, 1985), where x is the turnover rate of occupied exchanger at $E_m=0$ mV, and γ is an energy barrier parameter. Other terms: $\alpha_o = 1 + [Na]_o/K_{Na} + ([Na]_o/K_{Na})^2 + 2([Na]_o/K_{Na})^3$; α_i correspondingly; $\beta = 1/[(K_{Na})^3 \cdot K_{Ca}]$, where K_{Na} and K_{Ca} are dissociation constants of ion binding sites of exchanger. Numerical values of constants: $\gamma = 0.35$ (Kimura, Miyamae, & Noma, 1987), $K_{Na} = 71$ mM, $K_{Ca} = 1.5$ μM, and $\kappa = 100$, all obtained from computations with eq. (6.3) to generate curves that provided satisfactory fits to experimental data of figs. 6c and 8c of Kimura, Miyamae, and Noma (1987) and figs. 4A and 7A of Miura and Kimura (1989). The scaling factor k', of 0.71 μA $\cdot$ μF^{-1} $\cdot$ mM^{-4}, was determined separately by using eq.(6.3) for two of the conditions of Kimura et al. (1987, fig. 3) – that is, with $[Ca]_o$ of 2.0 and 0.1 mM, and setting equal to 1.66 μA $\cdot$ μF^{-1} (taken from third column of table 1 of the same reference) the difference between the two computed figures of i_{ex} obtained at $E_m = 0$ mV. Figure for $[Na]_i$ of 13.3 mM (range, 11–15 mM in other trabeculae) for computation of Q_{ex} [in (D)] obtained from trial and error runs made with eq.(6.3) until attainment of steady Q_m at time 100 ms $< t <$ 300 ms, compatible with complete inactivation of calcium conductance at this time (Campbell, Giles, & Shibata, 1988). Eq. (6.4) from Hodgkin and Nunn (1987, eq. 12c):

$$i_{ex} = \frac{k''([Na]_i^3[Ca]_o\exp(\gamma E_m F/RT) - [Na]_o^3[Ca]_i\exp(-(1-\gamma)E_m F/RT))}{(1+[Ca]_o/K_{Ca}+([Na]_o/K_{Na})^3)(1+[Ca]_i/K_{Ca}+([Na]_i/K_{Na})^3)} \quad (6.4)$$

where k'' is a scaling factor and γ, K_{Na}, and K_{Ca} have the same meaning as in eq. (6.3). Numerical values of constants: $\gamma = 0.35$ (see above), $K_{Na} = 18$ mM and $K_{Ca} = 6.7$ μM obtained with same procedure as for eq. (6.3). [N.B. The fit of curves constructed by means of eq. (6.4) was satisfactory for figs. 6c and 8c of Kimura et al (1987) and fig. 4A of Miura and Kimura (1989) but less so for fig. 7A of Miura and Kimura (1989). The value for k'', of 0.19 μA $\cdot$ μF^{-1} $\cdot$ mM^{-4}, was obtained in the same way as that for k' [eq.(6.3)].

worked out by means of eq. (6.3) (adapted from Mullins, 1977), from the action potential trace (panel A), the curve of changing $[Ca^{2+}]_i$ (panel B) and the fixed concentrations of $[Na]_o$, $[Na]_i$, and $[Ca]_o$ of 117, 13.3, and 2 mM, respectively (for $[Na]_i$, see legend). Integration of this current with respect to time then yielded the calcium uptake through the exchanger, Q_{ex} (panel D), and also Q_m, the calcium uptake through the calcium channel (panel E), by subtracting Q_{ex} from Q_c.

For a check of this approach, an alternative equation was used, eq. (6.4) (from Hodgkin & Nunn, 1987, eq. 12c), which, though derived with quite different assumptions from those of Mullins, nevertheless gave similar results, as the comparison of panels E and F shows.

The upshot of Fig. 6.10 is that at the peak of the Q_c curves (panels E & F), a substantial portion of the calcium input – some 33–38% of the total [i.e., $(Q_c - Q_m)/Q_c$] – seems to pass through the exchanger during each action potential (31–62% in similar computations with five other trabeculae). Since calcium activating tension in frog heart is derived mainly from influx through the cell surface (in the present conditions of low SR calcium discharge; Niedergerke & Page, 1989), agents affecting the exchanger flux like $[Na]_i$ should, therefore, alter contractile tension too, as suggested some time ago (e.g., Baker et al., 1969; Glitsch, Reuter, & Scholz, 1970). These and related ideas that were subsequently advanced (Mullins, 1977) can be summarised by saying that a rise in $[Na]_i$ would increase calcium influx through the exchanger during the initial part of the action potential while reducing calcium efflux during its later part, with a time course over a series of twitches depending on the rate of change of $[Na]_i$, which is likely to be slow. In the meantime, this mechanism of a $[Na]_i$-dependent calcium influx has received much support – for example, from the increased size of both aequorin light signal during inhibition of the Na–K exchange pump (Allen & Blinks, 1978), and exchanger outward current in sodium-loaded cells (Hume & Uehara, 1986a,b; for results with sheep Purkinje fibres, see Eisner, Lederer, & Vaughan Jones, 1983, 1984). Since $[Na]_i$ usually rises when the heart rate is increased (cf. Gadsby, Niedergerke & Page, 1971), this mechanism probably contributes to the staircase, together with other factors (e.g., the enhancement of cellular calcium uptake per unit time, see below; for a report of increased calcium inward current in this condition, see Noble & Shimoni, 1981).

However, explanations along these lines, by themselves, do not readily account for the effects of enhanced $[Ca]_o$ and reduced $[Na]_o$ which are not invariably, if at all, associated with increases of $[Na]_i$ [for reports of sodium decline in heart cells, resting or beating, at low $[Na]_o$ or enhanced $[Ca]_o$, see Ellis, 1977, and Deitmer & Ellis, 1978 (sheep Purkinje fibre); Sheu & Fozzard, 1982 (sheep ventricle); Chapman, Coray, & McGuigan, 1983 (ferret ventricle); Campbell & Tunstall, 1984 (frog heart atrium)]. More likely, the rapid increases of calcium influx evoked by high $[Ca]_o$ or low $[Na]_o$ give rise to a slower accumulation of calcium inside the cells (cf. Niedergerke et al., 1976), which, in turn, causes a slow twitch increase, events reflected in the biphasic responses obtained (e.g. Fig. 6.2, this chapter; fig. 16, Chapman & Niedergerke, 1970a). A similar sequence probably occurs after inhibition of the Na–K exchange pump or the increase in heart rate (staircase), even though the time course of these responses does not usually include a fast phase, presumably because of the slowness with which $[Na]_i$ and, thus, calcium influx, increase.

In addition to sodium, cytoplasmic calcium is known to affect calcium influx through the exchanger by acting at its "catalytic" site (DiPolo, 1979), an effect much enhanced by ATP (DiPolo & Beaugé, 1987). However, it is uncertain whether this mechanism contributes much to the processes under consideration in view of the low $[Ca^{2+}]_i$ at which it occurs [i.e., <100 nM, from estimates of half-maximal activation of calcium influx by $[Ca^{2+}]_i$ in guinea pig myocytes of 47 nM (Noda, Shepherd, & Gadsby, 1988) and 22 nM (Miura & Kimura, 1989)], that is, close to or below the lowest diastolic $[Ca^{2+}]_i$ so far recorded ($\geq 1 \times 10^{-7}$ M, Marban et al., 1987; Cheung et al., 1989; see, however, Frampton et al., 1990). Admittedly, this qualification is not decisive since (1) half-maximum $[Ca^{2+}]_i$ for this action is not yet known in the frog heart (being 1.8 μM & 0.3–0.7 μM in squid axon and barnacle muscle fibres, respectively; see DiPolo & Beaugé, 1987; Rasgado-Flores, Santiago, & Blaustein, 1989), and (2) the figures obtained for guinea pig heart cells may be underestimates, owing to the calcium buffers (EDTA & EGTA) which had been present, by necessity, in the determinations concerned (for a discussion on this point, see Reeves & Philipson, 1989).

Turning to the putative modifier action at sites on the myofibrils [factor Y of eq. (6.2)], recent results strongly suggest that in skeletal muscle phenomena like staircases and posttetanic potentiation are due to calcium (calmodulin)-mediated phosphorylation of the myosin light chain-2 (cf. Metzger, Greaser, & Moss, 1989; Sweeney & Stull, 1990); and it is of interest, therefore, to inquire whether this applies to the heart as well. Indeed, it is known that the effects of the kinase involved are similar in skinned heart and skeletal muscle fibres both in stimulating myosin phosphorylation and in enhancing contraction (Sweeney & Stull, 1986), but in intact cells the enzyme activity is more than tenfold lower in heart than skeletal muscle, because of the much lower kinase content of the former (Frearson, Focant, & Perry, 1976; Stull et al., 1982). Although for this reason myosin light-chain kinase is believed to play only a minor part in regulating the heart beat (Herring & England, 1986; Sweeney & Stull, 1986), this conclusion may not be final, as the kinase content of the frog heart is unknown and the proportion of phosphorylated myosin is said to be high in this tissue (i.e., 50%; Bárány et al., 1983).

The possibility of an alternative mechanism operating at the level of the myofibril emerges from x-ray diffraction studies of the heart by Matsubara, Yagi, and Endoh (1979; for a review, see Matsubara, 1980). By recording the intensities of the equatorial x-ray reflections to monitor the location of cross-bridges within the myofilament lattice, these authors found a marked difference between movements of cross-bridges into the "activating" position (i.e., the vicinity of the I-filaments) during systole and their return to the original position during diastole; the one occurred at a rate not dissimilar to that of tension development, the other much more slowly than relaxation. Thus, at the end of the diastole of a 1-s cardiac cycle, some 30% of the cross-bridges did not return to the position they occupy during quiescence; these cross-bridges are referred to as weak, as distinct from those generating tension. Since during paired pulse stimulation (and during staircases) the number of both weak and strong cross-bridges was found to increase together with the strength of contraction, a sequential mechanism was postulated in which cross-bridges remaining (and accumulating) in the "activating" position during diastole facilitate the formation of strong ones during systole, thereby increasing the heartbeat.

Conclusions

Slow hypodynamic twitch increases occur when either calcium influx through the Na–Ca exchanger (1), inward current through the calcium channel (2), or both (3) are enhanced [for (1) see Figs. 6.5, 6.6, & 6.7; for (2) see fig. 4 of Niedergerke and Page (1977); for (3) see, e.g., Figs. 6.1, 6.8, & 6.9]. A common factor in these conditions is the increase of cellular calcium, and it is likely therefore that these twitch changes are related to calcium-dependent processes (e.g., Niedergerke et al., 1976). In a reexamination of this idea (Figs. 6.3–6.6), it was found that processes related to the filling of cellular calcium sinks and of the SR cannot, by themselves, explain these changes (Figs. 6.7 & 6.8), suggesting that at least one other process is involved. An indication of its identity may reside in two properties attributed to the weak cross-bridges in cardiac cells (Matsubara, 1980): (1) a "priming" action through which they are able to increase the subsequent twitch, and (2) the low calcium concentration at which they are being formed (Matsubara et al., 1989). These features are in accordance with both the sensitivity change of preparations to calcium during the slow twitch changes (Fig. 6.9) and the "priming" effect ascribed to a portion of cell calcium accumulating (or subsiding) in association with these changes (Niedergerke, Page, & Talbot, 1969; Niedergerke et al., 1976). How is this calcium effect being conveyed to the myofibrils? There are at least two possibilities – the first including an intermediate step (e.g., activation of myosin light-chain kinase), the second acting more directly through calcium that occupies sites on one of the myofibrillar proteins (e.g., troponin). In both, an altered state of the contractile system so induced during a heartbeat, including its diastole, serves to amplify the subsequent systole. [As this state could involve "subthreshold" calcium binding to troponin, it is worth noting that during diastole up to about 10% of troponin-specific sites may be occupied by calcium, at diastolic $[Ca^{2+}]_i$ taken to be 3×10^{-7} M (Harding et al., 1989) and with K_D for these sites of 2.5×10^{-6} M (cf. Niedergerke & Page, 1989).] In this altered state, additional (weak) cross-bridges may be recruited, as implied by Matsubara (1980), but another, perhaps additional, effect controlling the rate of cross-bridge cycling is also feasible (Brenner, 1988a).

7

Regulation of contractile proteins in heart muscle

SAUL WINEGRAD

Introduction

Although it is often difficult and generally dangerous to try to identify the best piece of work or the most influential observation or the seminal result in a field of research, it is tempting to do so on this occasion. In order to make the task easier, I would prefer to list not one but a small group of contributions to the field of muscle biology, each devoted to a different aspect of the biology of the muscle cell. Few would question that A. V. Hill's (1938) paper on the energetics of muscle contraction belongs to this group of distinguished papers, and the separate but inter-related efforts of Ebashi and Weber to show that calcium ions were the activators of the contractile system have been equally important (Weber & Winicur, 1961; Ebashi, Ebashi, & Kodama, 1967). The introduction of electron microscopy to the study of muscle ultrastructure by H. E. Huxley in 1957 demonstrated the existence of two sets of interdigitating filaments and provided the ultrastructural basis for the basic mechanism of contraction. The remaining three developments that belong to this brief list of milestones to the understanding of muscle function were contributions of A. F. Huxley: (1) the constancy of the width of the A-bands over a wide range of lengths of striated muscle (A. F. Huxley & Niedergerke, 1954, 1958), (2) the local activation of a very limited por-tion of a single muscle fibre by the localized depolarisation of the surface membrane (A. F. Huxley & Taylor, 1955b, 1958), and (3) the first com-prehensive and quantitative theoretical model for the mechanism of contraction based on the existence of a large population of force genera-tors (A. F. Huxley, 1957a). All of this work, completed by 1961, defined the intellectual environment in muscle biology at the time I became a postdoctoral fellow with Andrew Huxley at University College, London, in 1961.

I had just completed some work showing that there was an increase in the influx of calcium ions into cardiac cells in association with electrically stimulated contractions, and that the extent of the increase was approximately linearly related to the amount of force produced by the contraction when it was varied by different frequences of stimulation (Winegrad & Shanes, 1961). In spite of the fact that the amount of calcium involved was small, approximately equal to one calcium ion for every hundred myosin molecules, the seductive powers of a linear relation were present and seemed to insist that calcium influx was a controlling, but still incompletely understood, step in the contractile cycle.

My original ambition for my postdoctoral period in London was to learn some of the techniques that were being used in the laboratory to study the myofibrils in the intact cell during the contraction, but Huxley's advice was different. The role of calcium in the activation of contraction was an important problem, and transmembrane fluxes could not be the complete story. He suggested that I should continue to work on calcium activation. The logical next step was to try to trace the intracellular movements of calcium quantitatively, to determine if a translocation of a sufficient amount from the sarcoplasmic reticulum to the myofibrils occurred to account for activation. During the discussions with Huxley that followed the decision to pursue this problem, I witnessed his extraordinary ability to clarify the salient features of an experimental project even when he had never used the necessary techniques or even worked in the area himself. I also learned of his wisdom in not telling all.

Before embarking on the project, it was quite clear that a theoretical analysis of the resolving power of calcium autoradiography, the only technique available at the time to use for following the translocation of intracellular calcium, was in order. I used a most favourable case scenerio and concluded that the technique had sufficient resolving power to answer the important questions. Huxley, in a more comprehensive analysis, had concluded that only in the most favourable model would autoradiography have sufficient resolution to follow critical movements of activator calcium. He also made the decision not to dilute my enthusiasm by communicating his more comprehensive, but less optimistic conclusions. Conviction that a problem is soluble can often carry an experimenter through difficult periods and even lead to interesting if unexpected results.

Although no useful data were produced during my year as a postdoctoral fellow in London, many of the important technical problems were identified and a few resolved. Over the subsequent eight years, it became clear that Nature had provided a best-case scenario for the calcium autoradiographer. It was apparent that calcium moved from the terminal cisternae of the sarcoplasmic reticulum to the region of the myofibril in which troponin was located and then returned to the reticulum at sites distributed along its entire length. Unexpected, however, was the time course, which showed that the calcium distribution within the cell at the end of a contraction was not the same as it had been before the contraction. It soon became clear that this slowness of recovery was one mechanism by which the muscle cell remembered the preceding mechanical events and by which activity influenced subsequent mechanical function.

At this point, I made a decision, again influenced by the examples and the subtle influence of Huxley's approach to research that I had carried with me from my time in England. Although a skeletal muscle cell is beautifully designed to perform a single task very well, its versatility over a brief time is limited. On the other hand, cardiac muscle cells must be capable of a broad range of function as regards development of force, velocity of shortening, and transduction of energy. Although, at the time, the field of cardiac cell physiology was focusing very heavily on modulation of contractility by alteration of calcium movements, it seemed reasonable to conclude that modification of the response of contractile proteins to activation by calcium ions was another important mechanism responsible for the broad range of function of which cardiac cells were capable.

Studies on cardiac muscle cells

In order to study the effect of changes in the contractile proteins on the variability of contractile function, it was necessary to use a preparation in which the functional properties of the contractile proteins could be probed with controlled concentrations of calcium ions. The effects of changes in excitation–contraction (e–c) coupling had to be avoided without disrupting the cell structure sufficiently to inactivate or destroy the regulatory reactions that were the object of the study. These requirements eliminated isolated myofibrils, glycerol-extracted fibres, and manually skinned cell preparations. Thomas (1960) had published a set of observations indicating that the sulphate space, normally restricted to the extracellular fluid in intact heart cells, expanded to include a volume equal to the intracellular space when cardiac tissue was treated with a high concentration of EDTA (ethylenediaminetetraacetic acid). This turned out to be an exciting lead because cardiac cells that had been treated with EDTA could be fully activated by the same low concentrations of calcium ions that activated isolated myofibrils and with a time course that was consistent with the absence of a major diffusion barrier at the cell surface. One property of this preparation, which has subsequently been exploited for other purposes in other laboratories, was the reversal of the high degree of membrane permeability when the EDTA was removed for a significant period of time (Winegrad, 1971). This reversal could be prevented by using EGTA instead of EDTA and by including the EGTA in a simple, simulated intracellular solution containing ATP but not magnesium ions. Although the EGTA-permeabilitised cardiac cell membrane appeared normal with conventional transmission electron microscopy, it was clear from the time course of force development that the cell was highly though probably not freely permeable to ATP, creatine phosphate, EGTA, and other small molecules and ions (McClellan & Winegrad, 1978). Functioning beta- and alpha-adrenergic and cholinergic receptors as well as adenylate cyclase remained in the membrane. The cells prepared by this treatment were called "hyperpermeable" muscle fibres (McClellan & Winegrad, 1978).

One of the first experiments that we carried out with the hyperpermeable fibres was rather naive, but the sort that sometimes stimulates the fantasy of an investigator. Rat hearts were homogenized, and then buffering systems for pH, calcium ions, and ATP were added to the homogenate. Hyperpermeable rat heart fibres were bathed in this modified homogenate. The rationale for the experiment was to see if there were any small molecules or ions in cardiac homogenate that would alter the contractile behaviour of the contractile system. The hyperpermeable membrane should exclude larger molecules and particulate components of the homogenate. The homogenate produced two changes in the relation between Ca concentration and force: (1) a decrease in the concentration of Ca required for both threshold and maximum activation of the contractile system, and (2) an increase in the maximum Ca-activated force. Both of these changes were reversible with withdrawal of the homogenate. Although changes of this general type can occur as a result of lowering the concentration of ATP, the changes that had been pro-

duced by the homogenate did not appear to have other characteristics of
the low-ATP response.

Of the two approaches we could have used to follow up these
observations – (1) fractionation of the homogenate to concentrate and
ultimately purify the molecule(s) responsible for producing the change,
of (2) making some guesses about what might be responsible and testing
the effect of the compounds in a simulated intracellular solution bathing
the hyperpermeable fibres – we chose the latter. Adrenergic and cholinergic agents as well as the two second messengers known at the time
(i.e., cAMP & cGMP) were tried. The two cyclic nucleotides had a rapid
and reversible effect on the Ca sensitivity of the contractile proteins.
Cyclic AMP shifted the tension–Ca relation to higher concentrations of
Ca, and cGMP produced the opposite effect. In order to make sure that
these compounds were not producing their effects on Ca sensitivity by
altering the permeability of the existing membrane to Ca or Ca-EGTA
in some way, we destroyed the sarcolemma of several preparations with
a low concentration of nonionic detergent after having first produced
shifts in Ca sensitivity with the cyclic nucleotides. The detergent fixed
the Ca sensitivity at its maximum level and eliminated the effects of
cyclic nucleotides on Ca sensitivity. A second effect was often also produced by detergent: maximum Ca-activated force was increased, sometimes by as much as 150–200%, but only when cAMP had been the last
drug applied before the detergent (Fig. 7.1). The results were inconsistent, however. Again, the influence of my time in Huxley's laboratory
and recollections of our occasional discussions provided useful guidance.
"Treasure your exceptions," a counsel from Huxley that I had taken to
heart so totally that I even find my wife and our daughters repeating the
advice and applying it to their own research efforts. At this time, investigators had begun to observe changes in the Ca sensitivity of isolated
cardiac contractile proteins in response to cAMP-dependent phosphorylation of troponin (Cole et al., 1978; Holroyde, Howe, & Solaro, 1979).
Although the extent of these changes was considerably less than we had
seen with cyclic nucleotides and hyperpermeable fibres, it was likely that
we were observing the effects of the same mechanism. While letting the
surprising effects of detergent incubate, we added radioactively labelled
ATP to our solutions and found changes in the degree of phosphorylation of the inhibitory subunit of troponin that very closely paralleled
the changes in Ca sensitivity of hyperpermeable fibres produced by
cyclic nucleotides. Beta-adrenergic agonists or cAMP produced phosphorylation of the inhibitory subunit of troponin (TnI) and a decrease in

Ca sensitivity. Cholinergic agonists or cGMP, on the other hand, produced dephosphorylation, and also an increase in Ca sensitivity, the extent of which depended on the amount of phosphorylation that had been reversed.

A pattern in the inconsistency of the increase in contractility following detergent treatment began to emerge with the recognition that the extent of change depended on (1) the length of time between the exposures to cAMP and to detergent, and (2) the age of the animal. The longer the delay after cAMP, the weaker was the response to detergent. In fact, the time-dependent decline was exponential, with a half-time of 4 min. If ATPγS was used as the phosphate donor, dephosphorylation occurred much more slowly, and the application of detergent could be delayed 60 min, with only about a 10% reduction in enhancement of contractility. Apparently, two reactions are involved in the increase in contractility: phosphorylation, and a second reaction produced by detergent. The effect of the age of the animal seemed to be related to the relative amounts of the two different isoforms of the heavy chain of myosin that were present: the greater the amount of alpha myosin heavy

Fig. 7.1. The change in calcium-activated force produced by treatment of hyperpermeable cells with cAMP, theophylline, and detergent. Numbers indicate pCa values.

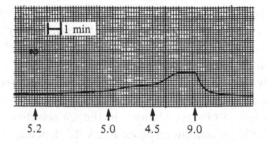

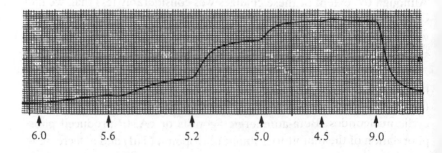

chain, the greater the response to cAMP–detergent. By this time several laboratories had already shown that the relative amount of the alpha isoform decreases with age in the rat heart.

Studies using radioactively labelled ATP during the cAMP–detergent treatment that increased contractility showed only two major phosphorylation sites on the contractile proteins: the inhibitory subunit of troponin (TnI) and C-protein. Work by England (1976), Garvey, Kramias, and Solaro (1988), and Silver, Buja, and Stull (1986), however, had shown that although phosphorylation of TnI could account for the shift in Ca sensitivity with beta-adrenergic stimulation, it could be dissociated from the change in contractility that also occurred. In light of these studies, the mechanism of the cAMP–detergent effect must involve either phosphorylation of C-protein or a protein other than the contractile proteins.

In order to produce larger amounts of hyperpermeable cardiac tissue for biochemical studies, we began to perfuse isolated hearts with EGTA. The resultant preparation was different from the hyperpermeable fibres produced by soaking trabeculae or papillary muscles in EGTA. No response to cAMP–detergent resulted, and electron microscopy revealed that the sarcolemma had large holes equivalent to about 45% of the total surface membrane. Reasoning that something critical to the (cAMP–detergent)-induced reactions might be extracted by the perfusion solution, we used a very small volume of perfused medium and recycled it. The result was the retention of the tissue's ability to respond to cAMP–detergent.

When the perfusion medium which appeared to extract a critical component of the contractility-modifying system was applied to other tissues with skinned cells, it produced an increase in maximum Ca-activated force. After concentration of the perfusion medium, and fractionation, the activity was contained in a fraction that produced only one major band, at 21 kD, on sodium dodecyl sulphate (SDS) polyacrylamide (PAGE) gels. The molecular weight was the same as that of the light chain of myosin, and the effect of the "active factor" on contractility could be reproduced with purified myosin light chains. The role of the regulating light chain seemed to be central, and our working hypothesis was that the change in contractility was due to movement of the light chain on and off the heavy chain, or modification of the light chain while it was bound to the heavy chain.

Throughout the studies on the hyperpermeable fibres, however, there had been one disturbing feature: the force developed by the preparations was relatively small, in the range of 4–20 mN/mm^2, whereas the maximum force that the cardiac contractile system could achieve, at least under certain circumstances, was about 100 mN/mm^2. Although values this low had often been reported in intact hearts, intact cells are rarely maximally activated with Ca. Hyperpermeable cells, especially after detergent treatment that would have removed any possible residual diffusion barrier to Ca, should be maximally activated. This raised some questions about the extent to which the phenomena demonstrated in hyperpermeable fibres were directly transferable to the intact heart.

To complement the studies of hyperpermeable fibres, in which force and velocity could be measured, we began to use the isolated working heart, in which many of the haemodynamic parameters, including work, preload, and afterload can be controlled. By quickly freezing the heart, it was possible to retain the regulatory state of the contractile proteins and at a later time, assay the performance of the contractile system by measuring the ATPase activity of myosin. With quantitative histocyto-chemistry, both Ca- and actin-activated ATPase activity of myosin could be determined, giving a measure, respectively of the hydrolytic activity of the enzyme site and the interaction of the cross-bridge with the actin filament (Weisberg et al., 1982; Winegrad & Weisberg, 1987). Since these measurements can be made in cryostatic sections of the frozen hearts that are about a fourth of the thickness of the ventricular cell, serial sections provide the opportunity for comparing the effects of different conditions on the myosin activity of the same cells. In essence, the cryostatic section consists of skinned cells in which the activity of the contractile proteins in a tissue that has already been characterised haemodynamically can be assayed by measuring ATPase activity. A further advantage of the preparation is the ability to measure the ATPase activity of V_1 and V_3 myosin isozymes in the same cells by exploiting the alkaline lability of V_3 and the alkaline stability of V_1.

In sections from both the quickly frozen nonperfused heart and the isolated perfused working heart, regulation of the contractile proteins has been shown in response to two different kinds of signals – beta-adrenergic stimulation and increased preload. Increase in the level of circulating catecholamine in the intact animal, addition of catechol-amine to the medium perfusing the isolated heart, and addition of beta-adrenergic agonist or cAMP to the solution bathing cryostatic sections of frozen hearts all cause an increase in both Ca- and actin-activated ATPase of myosin (Fig. 7.2; Winegrad, & Weisberg, 1987; Winegrad et al., 1986). The extent to which they are increased is proportional to the relative content of V_1. ATPase activity of V_3 is not increased. These data are quite consistent with the results of studies of the hyperpermeable fibres in positive or up-regulation of the contractile system from beta-adrenergic stimulation in proportion to the amount of V_1 present.

The effect of increased preload is seen only on hearts containing a mixture of V_1 and V_3 isozymes (Vato, Weisberg, & Winegrad, 1991). As the filling pressure is increased from 5 to 15 cmH$_2$O, cardiac work increases substantially and total ATPase activity remains approximately constant. The relative contributions of the V_1 and V_3 isozymes to the total ATPase activity, however, changes markedly. At 5 cmH$_2$O filling pressure in hearts containing about 60–65% alpha myosin heavy chain (which forms V_1), ATPase activity is due primarily to V_1. As preload increases, the contribution of V_3 increases and that of V_1 decreases. At 15–20 cmH$_2$O filling pressure, most of the myosin ATPase activity is due to V_3. Because the inherent ATPase activity of V_3 is only about 30% of an equal concentration of active V_1 molecules, replacement of V_1 by V_3 with the same total ATPase activity means than that there are substantially more enzymatically active myosin molecules at 15–20 cmH$_2$O than at 5 cmH$_2$O filling pressure. It seems reasonable to assume that this phenomenon, which results in an increase in the number of active force generators as preload rises, contributes to the length–tension or pressure–volume relation originally described by Starling many years ago.

In hearts from young rats, in which the isoform of myosin is all or nearly all V_1, changes in preload do not alter the ATPase activity. Although there is as yet no conclusive evidence for this hypothesis, the rat heart cell's response to preload seems to involve a mechanism that up-regulates V_3 with increasing preload, and in some mechanism as yet not understood, the up-regulated V_3 is favoured during muscle activation. In the absence of V_3, there is no change in the response of V_1 to Ca activation.

Fig. 7.2. The effect of 1 μM cAMP on the myosin ATPase activity in serial cryostatic sections of rat left ventricle: (A) control; (B) 1 μM cAMP. ATPase activity is proportional to optical density.

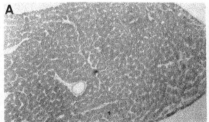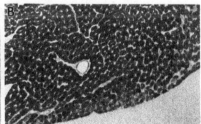

There is an interaction between the effects of beta-adrenergic stimulation and preload on the activity of V_1. As increasing preload downregulates V_1, it also decreases the ability of beta-adrenergic agonists or cAMP to up-regulate V_1. This inhibitory effect of increasing preload is progressive, becoming stronger as the preload increases, and at 20 cmH_2O filling pressure, 1 μM cAMP has very little effect on V_1.

Throughout these studies of regulation of contractile proteins and those of other laboratories of membrane channels, and the sarcoplasmic reticulum, an investigator becomes increasingly aware of the sophistication of function of the cardiac contractile process. The mechanisms by which the heart adapts to changing conditions to maintain its viability and still perform its necessary functions for the survival of the organism become progressively more evident. Another of Huxley's sayings is germane: "You needn't take credit because Nature is so clever." If the experimenter asks the right question, and employs a suitable technique to seek the answer, Nature will open its door a trifle wider, and for a brief time, he or she will feel a sense of greater understanding. However, as Pascal realised when he compared knowledge to a sphere, what you see increases as the square of the radius, but what you do not see increases as the cube.

8

Differential activation of myofibrils during fatigue in twitch skeletal muscle fibres of the frog

HUGO GONZÁLEZ-SERRATOS AND
MARIA DEL CARMEN GARCIA

Introduction

I spent three wonderful years in Professor Sir Andrew Huxley's laboratory at University College, London. When I arrived, I was very surprised and glad to find that he not only could understand my broken and poorly pronounced English, but that he was always extremely patient and calm with my ignorance, and with my naive questions. Thus, very soon he made me feel as if I were at home in a warm, friendly, and productive atmosphere. His kindness, superb knowledge, intelligence, and memory were continuously evident as a marvelous deep fountain from which to learn and acquire education. This education went beyond the walls of the laboratory into many areas of life and was soon complemented by the deep mark that Richenda Huxley also made on my outlook and education. I will never forget one day when Professor Huxley and I were discussing some of my experiments on isotonic recorded contractures, he very soon said something like, "Well, if you have an isotonic lever with such a spring and friction, then the sarcomere length that will be reached during the contracture would be . . . 2.4 μm". It took him around two seconds to figure out that sarcomere length! After he left I measured the sarcomere length reached during one of such

We thank Dr. G. McClellan for his help in the preparation of the manuscript. This work was supported by NIH grant R01 NS17048. M. C. Garcia is a trainee of CONACYT (Mexico).

experiments, and it was 2.41 μm! It took me three days to arrive at the same figure he computed in those few seconds in his brain. As we used to say in the laboratory, "His brain goes click, click, click, and the right answer from these elaborate calculations comes out within seconds". During my stay in his laboratory, I worked on three problems: all three took advantage of the formation or disappearance of wavy myofibrils and the use of cine-microphotography or still microphotography. One of the problems concerned the differential activation of myofibrils during potassium contractures. Another involved measuring the velocity of inward spread of activation of the excitatory process along the T-system in order to understand the nature of that signal. The third, on which Lucy Brown did most of the work, was to investigate the mechanism by which the myofibrils become wavy. The work presented here utilised similar techniques and is based on the observation of formation of wavy myofibrils during muscle fatigue.

Muscle fatigue

Prolonged, direct electric stimulation of vertebrate striated muscles induces a state (hitherto referred to as fatigue) during which the contractile force first declines, and subsequently, the muscle becomes mechanically refractory to further stimulation. This state involves neither permanent impairment of function, since contractility can be restored after an adequate period of rest (Westerblad & Lännergren, 1986), nor neuromuscular transmission failure, since it can be produced in the presence and absence of curare. Bigland-Ritchie and Woods (1984) found that the reduction of force developed during prolonged voluntary contraction (i.e., muscle fatigue) is due neither to decline in central nervous system motor drive nor to failing of neuromuscular transmission, but solely to contractile failure of the muscle involved.

The decrease in contractile force that occurs during fatigue is not well understood. There are two main trends of thought as to the mechanism of muscle fatigue: that it involves (1) alterations of the ATPase system (metabolic proposition), or (2) a failure of one or more links in the chain of events that couple excitation with contraction (e–c coupling proposition). Regarding the metabolic alterations that occur in muscle fatigue, various proposals have been offered: (1) an increase in ATPase activity, which leads to a reduced concentration of phosphocreatine and ATP (Spande & Schottelius, 1970; Edwards, Hill, & Jones, 1975; Dawson, Gadian & Wilkie, 1978, 1980); (2) an accumulation of ADP, orthophosphate, lactate, and H^+ (Fitts & Holloszy, 1976; Wilkie, 1981; for a review of these topics see Porter & Whelan, 1981); and/or (3) a change in the actin–myosin interaction (Dawson et al., 1978). The products of the ATPase activity (ADP, P_i, & H^+) might, after prolonged stimulation, impair the essential biochemical process of contraction, or the decrease in intracellular pH might interact with the actomyosin complex (Edman & Mattiazi, 1981; Mainwood & Renaud, 1985) leading to mechanical fatigue. However, a direct causal relationship between the observed metabolic changes and the development of fatigue has not been clearly demonstrated (Fitts & Holloszy, 1976; Edwards, 1981). Wilkie (1981) suggested that alterations in phosphocreatine (PC), P_i, MgATP, and H^+ do not distinguish between the two main propositions. At present, there is evidence that supports the second proposition: that is, that fatigue is probably due to a failure of one or more of the e–c coupling chain of events (Eberstein & Sandow, 1963; Grabowski, Lobsiger, & Lüttgau, 1972; Vergara, Rapoport, & Nassar-Gentina, 1977; Gonzalez-Serratos et al., 1978; Nassar-Gentina, Passonneau, & Rapoport, 1981; Bianchi & Narayan, 1982). Changes of resting membrane potential and action potentials cannot account for fatigue (Eberstein & Sandow, 1963; Grabowski et al., 1972; Lännergren & Westerblad, 1987). Nor can it be explained as due to failure of the contractility machinery to produce force (Eberstein & Sandow, 1963; Grabowski et al., 1972; Gonzalez-Serratos et al., 1978). Neither is it due to an exhaustion of energy resources (Grabowski et al., 1972; Vergara et al., 1977) since there is no depletion of glycogen or ATP (Nassar-Gentina et al., 1981).

Gonzalez-Serratos et al. (1978) and Somlyo et al. (1978) found that in fatigued muscle cells the T-system was swollen, forming vacuoles; the calcium concentration in the terminal cisternae (TC) of the sarcoplasmic reticulum (SR) was higher than the normal resting value; the mitochondria were not swollen, and their Ca content was low; and the total cytoplasmic calcium was slightly increased. The occurrence of vacuolation in the T-system raises the possibility that the tubular action potential fails to travel along the T-system to reach the centre of the fibre, or that a failure of the signal between the T-system to the TC in the swollen T-system regions fails to bring about the release of Ca^{2+} from the TC. Either one of these possibilities would lead to a decrease in the number of myofibrils activated, while others would remain active and would lead, therefore, to decrease in the force development during the mechanical activation.

If, on the other hand, the decrease in tension that appears in fatigue is due to a generalised process such as changes in the energetic sources, decreased intracellular pH, alteration in the interaction between the contractile proteins or dysfunction of the SR, it would be expected that all the myofibrils would be affected equally, gradually leading to a lower activation of all of them.

We investigated the above possibilities by detecting whether myofibrils were differentially activated or not. Our approach was to observe whether during fatigue, when a muscle cell was allowed to shorten, some activated myofibrils would remain straight, whereas others would not be activated, becoming therefore wavy. The same principle had been used in the past to distinguish between active and less active or inactive myofibrils (e.g., A F. Huxley & Gordon, 1962; Taylor & Rüdel, 1970; Gonzalez-Serratos, 1971, 1975; Bezanilla et al., 1972). We found that the development of fatigue was accompanied by the progressive appearance of wavy myofibrils, and that exposure to 20 mM caffeine straightened them up at the same time that tension recovered to nearly 100% of the tetanic tension. We concluded, therefore, that fatigue might be due to a failure of either the tubular action potential or the T-system to the TC signal. Preliminary reports of these results have been presented elsewhere (Gonzalez-Serratos et al., 1981; Garcia & Gonzalez-Serratos, 1982).

Methods

If fatigue is due to an interruption of the tubular action potential or of the tubular membrane to TC signal, myofibrils in the area of interruption would not be activated. However, if fatigue is due to a general metabolic alteration that directly affects the myofibrils, then no differential activation of myofibrils should occur, and all the myofibrils in the cross-sectional area of the cell should have been equally less activated.

The principle of the method consisted then in ascertaining whether during fatigue all the myofibrils were activated (to a given degree), or whether some myofibrils were not activated as compared with the control tetanic contractions. To distinguish activated from nonactivated myofibrils, some of the tetanic contractions elicited before and during the development of fatigue were recorded isotonically, whereas the rest were recorded isometrically. During the isotonic contractions, the shortened activated myofibrils remained straight, whereas the nonactivated ones folded and became wavy, once they reached an average sarcomere length ($\bar{s}$) of 2 μm or less (A. F. Huxley & Gordon, 1962; Brown, Gonzalez-Serratos, & Huxley, 1984a,b).

Single twitch fibres isolated from either the semitendinosus or tibialis anterior of the frog (*Rana temporaria*) were mounted in a narrow chamber with one tendon attached to an RCA 5734 mechanoelectric transducer for recording tension, and the other to an isotonic lever for shortening steps. During isometric contractions, the fibres were held at an average sarcomere length of 2.2–2.4 μm. The amount of shortening required to decrease the average sarcomere space from 2.4 μm to 1.7 μm was predetermined. By limiting the decrease in $\bar{s}$ to 1.7–1.8 μm, inactivation due to shortening below 1.5 μm (Taylor & Rüdel, 1970) was avoided. The presence of active or nonactive myofibrils was determined with cine-microphotography. The optical system used has been described previously (Bezanilla et al., 1972; Gonzalez-Serratos, 1975).

Results

Fatigue might be expected to affect contractility, longitudinally as well as radially. If there were a longitudinal effect, some portions of the fibre would contract less than others. This would lead then to an increase of the series elastic element and therefore to a smaller and slower development of tension. The first series of experiments were designed to determine whether during fatigue there is uniform shortening of the muscle cell. In these experiments the cine-micrographic recordings were taken with a 50-mm low-power objective (N.A. 0.08) and a ×4 eyepiece. The movement of markers attached along the cell was followed during the tetanic stimulation both during the early force development and during the shortening, after the release. The results showed that before and after fatigue, the shortening was uniform and in the same proportion from the fixed tendon to the moving one. Therefore, development of fatigue in the longitudinal axis is homogeneous.

We next examined the mechanical activation of individual or groups of myofibrils during tetanic stimulation with release and shortening to bring $\bar{s}$ close to 1.7 μm before and after fatigue. Figure 8.1 shows the effect of shortening of the muscle fibre to a sarcomere length of 1.6 μm $< \bar{s} <$ 1.8 μm on the behavior of myofibrils before and after fatigue. The sample micrograph (B-a) and force level (A-a) were taken after 29 tetanic stimulations when the tetanic force was 99% of the control. Neither here nor in any of the 15 experiments of this type performed was there any indication of wavy myofibril formation at a longer or shorter

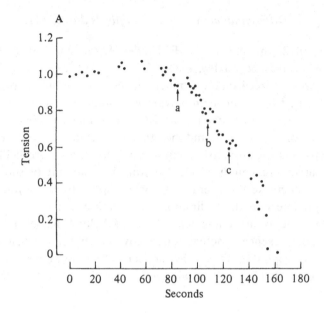

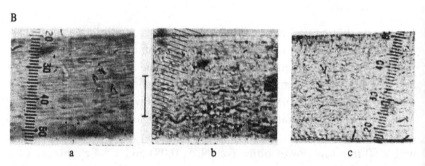

Fig. 8.1. Time course of the development of fatigue and the formation of wavy myofibrils. (A) Time course of tetanic tension changes during a prolonged repetitive stimulation. Tetanic tension is plotted as a fractional change of the control tetanic tension. Fatigue was established by repetitive stimulations of three times the voltage threshold at 70 Hz for approximately 300 ms, followed 1.2 s later, by a single stimulation. This cycle was repeated every 3 s. At different times during and after the development of fatigue, cine-microphotographic recordings were taken during 800-ms tetanic stimulation. During the shortening due to releases of the isotonic lever to decrease s̄, the degree of movement of the lever was controlled by two stops – one behind it pressed against the lever, the other in front of it separated by a given distance to permit a predetermined amount of shortening. Both could be adjusted with micrometer screws. A third stop, pressed against the front part of the lever, prevented it from moving until it was quickly released during the mechanical activation. (A) a, b, and c indicate the tensions at which the corresponding sample micrographs were taken. (B) Sample micrographs with cine-micrographic recordings taken at 65 frames per second at times A-a, A-b, and A-c. Calibration bar in (B) is 50 μm for b and 52 μm for a and c. Temperature, 21°C.

fibre length. In contrast, when the force development had dropped to below 90% (A-b) of the control tension, although no wavy myofibrils were seen before the release ($\bar{s} \simeq 2.2$ μm), they appear very clearly after it, as shown in Fig. 8.1 B-b. The micrograph (B-c) shows that when tension development dropped to around 56% of the control, the number of wavy myofibrils increased. In all similar experiments wavy myofibrils were never observed before fatigue developed, but they did appear clearly once fatigue started, provided the fibre was allowed to shorten to a new sarcomere length of 1.6 μm $< \bar{s} <$ 1.9 μm. The only time wavy myofibrils were observed *before* fatigue was established was when there was a clear indication in the tension recording of a probable failure of either the sarcolemmal or perhaps the tubular action potential (Bezanilla et al., 1972). In that case, as shown, the shape of the force development at longer $\bar{s}$ was similar to those illustrated in reports published elsewhere (Bezanilla et al., 1972).

The finding that fatigue and wavy myofibrils appeared at the same time indicates to us that these wavy myofibrils were either not activated or activated less than the straight ones, as had been proposed by A. F. Huxley and Gordon (1962), Taylor and Rüdel (1970), Gonzalez-Serratos (1971, 1975), and Bezanilla et al. (1972). This differential activation could be due to either a lower Ca^{2+} release from the TC or a reduced Ca^{2+} sensitivity of the myofibrils during fatigue. This possibility prompted us to investigate the effect of caffeine on fibres that had developed wavy myofibrils. A caffeine-induced activation would be expected to release the Ca^{2+} from the TC of the wavy myofibrils and straighten them up owing to their activation, and to bring about a force development similar to the control. As illustrated in Fig. 8.2A-a and B-a, before fatigue there was no indication of wavy myofibril formation after the release during the tetanic stimulation. But, as shown in Fig. 8.2A-b and B-b, once fatigue developed, there was a clear appearance of wavy myofibrils at a short $\bar{s}$. Furthermore, Fig. 8.2A-c and B-c shows that when the muscle cell was superfused with a Ringer's solution containing 15 mM caffeine, the wavy myofibrils straightened up and the force developed became similar to the control. It was also found that with lower caffeine concentrations (1–5 mM), the wavy myofibrils straightened up less and the recovery of force was smaller.

Discussion

These results show very directly that fibres allowed to shorten to 1.6 μm $< \bar{s} <$ 1.9 μm during fatigue develop wavy myofibrils which are taken as a sign that they either have not been activated or are less activated than the straight ones and that when exposed to enough caffeine the wavy myofibrils become straight, and tension similar to the control can be developed. These results lead us to conclude that the drop in tetanic tension development during fatigue is due primarily to the failure of myofibrils to be activated during electrical stimulation. That myofibrils

Fig. 8.2. Effect of caffeine on wavy myofibrils and tension that developed after fatigue was established. (A) Sample micrograph of cine-recordings taken during the mechanical activation after the release: (a,b) during the tetanic stimulation; (c) during exposure to 15 mM caffeine. In (B), the corresponding tension recordings are shown. The decrease in tension during tetanic stimulations and the caffeine exposure are due to the shortening of the cell after the release of the isotonic lever. Time calibration bar: 24 s for a and b; 12 s for c. Temperature: 22°C.

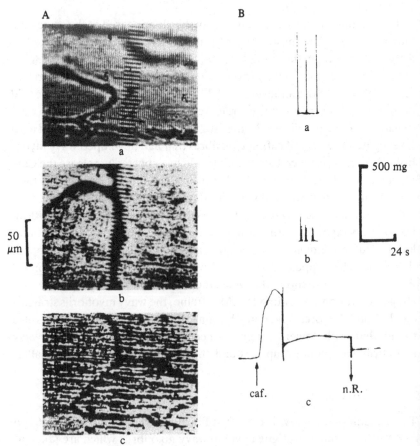

are capable of independent contraction has been shown previously by several researchers (A. F. Huxley & Gordon, 1962; Taylor & Rüdel, 1970; Gonzalez-Serratos, 1971, 1975; Bezanilla et al., 1972). Since the wavy myofibrils are not activated, the most probably explanation is that in these myofibrils the Ca^{2+} inside the TC is not completely released or not released at all. Gonzalez-Serratos et al. (1978) and Somlyo et al. (1978) have shown that the calcium content of the TC in fatigued fibres remains as high as, or higher than normal so that when caffeine is applied, this Ca^{2+} is released, the myofibrils are activated and straightened up, and tension is also recovered.

The increase in the proportion of inactive myofibrils as fatigue progresses could partially account for the observed decrease in velocity of shortening observed by Edman and Mattiazi (1981), since these myofibrils would represent an increased internal load against which the active myofibrils must contract during shortening.

The mechanism of fatigue remains to be established, although the present study, together with the one published previously (Gonzalez-Serratos et al., 1978; Somlyo et al., 1978), clearly shows that it is not due to depletion of Ca from the terminal cisternae. Other mechanisms that might contribute to fatigue include:

1. *Decreased affinity of a binding site (i.e., troponin) for Ca* (Fuchs, Reddy, & Briggs, 1970; Fabiato & Fabiato, 1978; Cooke et al., 1988) due to lowering of intracellular pH (Gonzalez-Serratos et al., 1978; Westerblad & Lännergren, 1988). This mechanism is unlikely since fatigued muscles can respond to caffeine, developing force after fatigue (Eberstein & Sandow, 1963; Garcia & Gonzalez-Serratos, 1982, and the present study), and since there is a lack of correlation between fatigue and a decrease in intracellular pH (Westerblad & Lännergren, 1988).

2. *Uncoupling due to mitochondrial Ca loading.* This mechanism is unlikely since the mitochondrial Ca content in fatigued muscles is relatively low and the structure of the mitochondrion is not altered (Gonzalez-Serratos et al., 1978; Somlyo et al., 1978).

3. *Failure of the tubular action potential secondary to their swelling* (Gonzalez-Serratos et al., 1978).

4. *Failure of the signal between the T-tubule membrane and TC in some regions of the T-system* (e.g., where the swelling occurs).

The persistent response of fatigued muscles to high K or caffeine, the high Ca content of the TC (Eberstein & Sandow, 1963; Gonzalez-Serratos et al., 1978), and the results presented in this chapter are compatible with the last two mechanisms. The failure of the tubular action potential is probably less likely since the observed changes in the sarcolemmal action potential are not sufficient to explain fatigue (Eberstein & Sandow, 1963; Grabowski et al., 1972; Lännergren & Westerblad, 1987). We therefore propose that the likely mechanism to explain the fatigue of the myofibrils to become activated is then a failure of the signal between the T-system and the TC.

9

High-speed digital imaging microscopy of isolated muscle cells

STUART R. TAYLOR AND KENNETH P. ROOS

Introduction

I remember Tuesday, February 25, 1964 as the day Andrew Huxley's boyhood hobby and long-standing interest in microscopy captivated my interest and thereafter strongly influenced my activities for the following 25 years. On that day, while I was still a graduate student, Andrew Huxley delivered the first of several lectures in the Jessup Lectureships at Columbia University. His topics included (1) "The Microscope Image of Striated Muscle", (2) "Changes in the Striation Pattern during Contraction", (3) "Past and Present Opinions on the Structure of the Muscle Fibre", (4) "The Length–Tension Diagram of Striated Muscle", (5) "The Inward Spread of Activation of Contraction", and (6) "The Steps Between Electrical Activity and Contraction". A leap of three and a half years brings us to August 29, 1967, the day I began postdoctoral training in Andrew Huxley's laboratory at University College; on that date, he lent me his copy of Barer's "Lecture Notes on the Use of the Microscope". In retrospect, these notes probably fixed an agenda that had first been proposed to me during those three weeks in 1964.

Compelling arguments for studying the contraction of fresh muscle

This work was supported by grants from the NSF (DMB-8503964, INT-85-16610, and INT-86-10707), the NIH (NS-22369), and the Pittsburgh Supercomputing Center (DCB-890009P) to S.R.T., and the NIH (HL-29671) and the UCLA Laubisch Endowment to K.P.R. We thank V. Arlene Morris for the primary cell cultures and criticism of the manuscript, Jon Charlesworth for the electron microscopy, and Mark J. Patton and Laura A. Quesenberry for hardware and software development.

fibres under the microscope had been made before and have been made since those lectures in 1964 (e.g., A. F. Huxley, 1977). However, light microscopy has found applications in many new areas of biological research in the past decade (Allen, 1986). Among the factors contributing to this renewed interest are the development of light-emitting probes that are sensitive and quantitative detectors of ions and cellular structures (e.g., Tsien, 1983; Lin, Jovanovic, & Dowben, 1989), the development of microscopes that can illuminate and record images from an area of a cell restricted to the plane of focus only (e.g., see Somlyo, 1986), and progress in developing computer programs that allow rapid, automatic extraction of subtle image information as well as precise control of a microscope (Inoue, 1986; Bradbury, 1989). Cell biologists in general, and particularly those who have closely followed the applications of microscopy to studies conducted in A. F. Huxley's laboratory, realize that there were several periods prior to the 1980s in which microscopy led to findings that were pivotal in the attempts to understand better the mechanisms of muscle contraction and excitation–contraction (e–c) coupling (A. F. Huxley, 1977).

The speed of events in muscle makes it pointless to apply computer programs that enhance contrast to standard video images of small, virtually invisible objects such as striations, when the data might be as little as one blurred frame during a twitch at room temperature. One productive alternative has been to generate contrast with the microscope, to project images of striations onto a linear array, and to monitor the time course of one-dimensional changes in striation spacing (Roos & Brady, 1982). However, an ordinary microscopic image of striations only, does not contain sufficient information to tell whether the dark and light bands correspond to A- or I-bands because the plane of focus is unknown (A. F. Huxley, 1957a, 1977). Consequently, digital imaging light microscopy has not been used on striated muscle to achieve insights about the contractile proteins actin and myosin that are comparable to studies of actin, dynein, kinesin, and tubulin which have aided understanding the biology of other cells (e.g., Allen, 1986; Inoue, 1986).

Light microscopy can continue to play an important role in future studies of striated muscle. However, unique problems must be solved before one can use microscopy routinely in conjunction with developments in digital image processing and pattern recognition. First among the physiological problems is the fact that a single cycle of contraction and relaxation involves several very rapid events that follow an extrinsic stimulus (Block et al., 1988; Rios & Pizarro, 1988). Hence automatic, high-speed detection, extraction, and description of multidimensional patterns of motion in living muscle cells currently require a custom-made computer vision system. Second, one must either use some method to measure the actual value of the refractive index in a given band of each striation (e.g., A. F. Huxley & Niedergerke, 1954) or simultaneously monitor (a) the position of the myofibrils containing visible bands or (b) some objects closely associated with visible bands and the positions of the fibrils, to deduce or compare contraction in adjacent sarcomeres (Brown, Gonzalez-Serratos, & Huxley, 1984a,b).

The development of compact, solid-state area cameras that can read out selectable lines of image information in sequence (EG&G Reticon, 1989), and the possibility of combining these cameras with processing hardware and software that can begin analyzing data before a frame is even acquired, is a promising way to apply light microscopy to the study of living muscle. However, there are also obstacles to obtaining a system capable of technology's maximum performance. Among these are the needs for custom-made data formatters, as well as extensive software development to control all the programmable parameters of image acquisition, enhancement, recognition, automatic quantitative analysis, and consequently, the very substantial expense in direct costs and time.

Our purposes here are (1) to summarize our personal experience with a method of combining optical microscopy with computer enhancement and object recognition, (2) to outline some of our observations during twitches of muscle cells containing sarcomeres, and (3) to compare some features that may be specialised in a particular type of muscle with features that may be intrinsic to thick–thin contractile filament interaction.

Previous findings with light microscopy

Striations and their organisation were long ago recognised as one of the very few features intrinsic to living muscle that one can use to study contraction, activation, and relaxation (A. F. Huxley, 1957a, 1977). When a living muscle cell is properly focussed in the light microscope, its striated pattern is invisible to the human eye or a photographic emulsion, and the most common procedures to permit the study of striations are to defocus the image slightly, or to add elements to the optical system that generate contrast among objects having a different refractive index (A. F. Huxley, 1957a, 1977).

In early 1954, Huxley and Niedergerke were the first to see both striations and wavy fibrils near the cathode of a slowly increasing current in local contractions of adult skeletal muscle fibres. They did not realise the significance at the time but later thought that the waviness might be an indication of passive shortening. This chance observation was clearly in mind when Huxley and Gordon looked at constant-current contractures again in 1962. Waviness implied that some long-range physical connection, perhaps a part of the cytoskeleton (Street, Sheridan, & Ramsey, 1966), tightly coupled actively shortening sarcomeres of skeletal muscle to those that were passively dragged back and forth. It also became clear from later studies in Huxley's laboratory that striation spacing and sarcomere length are not necessarily equal and that these expressions of molecular length cannot always be used interchangeably. The quantitative difference between the two can be substantial under some circumstances, and, at least in adult skeletal muscle, one must know both the striation spacing and the orientation of the myofibrils in which the striations reside to deduce the associated sarcomere length.

The orientation of the myofibrils can sometimes be inferred from the position of associated objects, such as nuclei and chains of mitochondria, which lie in clefts between myofibrils and help to outline a fibril's position (e.g., Costantin & Taylor, 1973). In addition, striations and the orientation of mechanically coupled myofibrils are the only features intrinsic to adult skeletal muscle that permit one to detect possible differences in the level of activation throughout a cell's volume (Brown et al., 1984 a,b).

The results of Huxley and Niedergerke's original experiments that received the greatest attention and led to the most provocative inferences in modern muscle physiology were the near constant length of the thick filaments and the likelihood that short-range physical connections (cross-bridges) between thick and thin filaments were fundamental features of muscle contraction. These things are usually described as part of the "sliding filament hypothesis" or the more inclusive "theory of force generation by independent cross-bridge interactions between myosin and actin" (A. F. Huxley, 1977).

However, their less well-studied observation of wavy fibrils has been used several times in microscopic studies of adult vertebrate skeletal muscle to infer spatial and temporal differences in activation (e.g., Brown et al., 1984a,b). Several years ago, Podolsky (1962) raised the possibility that Starling's law of the heart was primarily a law of activation rather than a law of muscle cells in general. It seemed that microscopic study of adult cardiac muscle cells might also show differences that would allow Podolsky's suggestion to be tested in the same way it was being tested on skeletal muscle. Unfortunately, active changes in the length of isolated cardiac myocytes do not produce myofibril buckling or differences that are readily detected by conventional microscopy and photography (Krueger, Forletti, & Wittenberg, 1980). However, contraction bands are not observed in shortened cardiac myocytes either. Hence these relatively easy-to-detect features in adult skeletal muscle cannot be used in cardiac cells to estimate possible regional differences in contraction.

Recent findings with microscopy and computer vision

We have made use of a system still in development known as CAMERA, an acronym for Computer-Assisted Measurements of Excitation–Response Activities. CAMERA permits each two-dimensional (2D) image of a microscopic portion of a cell to be catpured, with exposure times ranging from nanoseconds up to a maximum time of about 4 milliseconds (ms). At least 40 full-size square frames to more than 1000 rectangular fields of sequential images can be exposed and read into digital memory at speeds much faster than images taken with standard broadcast formats. The images can be synchronised with precisely timed extrinsic stimuli, and 2D and 3D mapping of muscle cell features can be followed and analysed in real physiological time (Roos et al., 1989a).

We wondered whether the previous failure to observe differences in regional shortening between cardiac and skeletal muscle cells might arise from differences outside the contractile filaments – for example, from a difference in the influence of long-range mechanical connexions in the cytoskeleton. The first reported use of the CAMERA system confirmed that wavy myofibrils and striations could be detected in living frog skeletal muscle, although these image details were subtle, and contrast was not enhanced by methods intrinsic to the microscope (Helland et al., 1988). Subsequently, this same system also allowed us to detect subtle differences in shortening in distinct areas of rat cardiac cells (Roos & Taylor, 1989; Roos et al., 1989b).

Methods

The optical system of CAMERA is as simple as possible. It consists of only a single microscopic objective connected through spring-loaded levers (A. F. Huxley, 1961) to a microstepper. The optical system contains no aperture stop, collimator, field lens, eyepiece, or factors other than room air to absorb light energy or introduce aberrations between the back surface of the objective lens and the quartz window of a pixel image sensor (MC9128 Camera; EG&G Reticon, 1989). The micro-stepper permits us to make precise changes in focus under remote control. The objective can be focussed stepwise without detectable oscillation in minimum step sizes of 0.5 micrometers (μm) at speeds up to at least 2 μm/ms.

Some of the unique features of CAMERA can be appreciated by a brief comparison of this system with a conventional, unmodified video camera configured for one of the standards primarily developed for television applications. Greater detail about the system can be found elsewhere (Roos et al., 1989a). The start and end of frame exposure can be controlled independently when we acquire data, whereas cameras with a standard interface continuously form images at a rate unrelated to external events. This is a disadvantage in cell physiology where one usually wants to synchronise image capture with an external stimulus.

We can acquire data rapidly as a matter of course, at rates from 232 full frames per second and partial frames at several thousand frames per second, whereas standard formats are limited to only a couple of frames per muscle twitch (30 Hz interlaced for composite video or 60-Hz noninterlaced analog signals generated by a workstation). Commercial providers of imaging workstations claim to acquire images with "real-time" digitation, when in fact they mean a video format that is 60 frames per second or slower. This rate is not capable, however, of acquiring images in the real-time of a physiological event such as a muscle twitch. We can routinely change the frame speed to exceed the rate at which aliassing is evident.

Although algorithms have been developed to calculate or measure and correct for some of the causes of image blurring, such as out-of-focus material or microscope objective distortion, they obviously cannot correct for images never captured owing to rapid movement of an object. A quantitative description of the image-forming power of a system can be derived either by analysis of the fineness and separation of points and lines in an object or by measuring and comparing the ratio of image to object with spatial frequency. Programs are available to analyse automatically the image-forming power of a system and deduce the size and extent of the focus blur. But the problem of determining the factors to correct for motion blur has not yet been solved. The CAMERA system was designed to reduce greatly the motion blur in the original data and then take advantage of ongoing developments in processing algorithms that deblur focus distortion in images to calculate automatically the speed and direction of a moving object.

Lines of our data are sequential and available one right after another. This allows us to use image processing algorithms that operate on contiguous bits of data and to start processing a frame before it has been completely exposed. In standard formats one frame is usually two overlapping fields (i.e., interlaced: odd numbered lines then even numbered lines, are read out), and the image processor must sit idle until both fields arrive. Our format is square, and an object covers the same number of pixels when rotated, which makes the calibration the same in all directions. Standard video usualy has an aspect ratio of 4/3, which requires that horizontal measurements have a different calibration than vertical measurements. Most typical cameras also have a fixed-duration horizontal synchronisation and blanking period overlaid onto each line of data, which increases the likelihood that aliasing can become a problem.

Several companies now market motion-recorder video systems that are integrated with PC compatible workstations, optical disc recorders, and software programs for automated data acquisition, shutter control, and image processing. They can digitise one field in 1/60th of a second at half resolution, and store it as a 24- to 32-bit data block. The current cost is about one-tenth the original cost of the 12-bit system we are using. But the performance of these workstations is not adequate for the rapid events in muscle contraction. These commercial systems are, however, useful for time-lapse light microscopy or for eliminating the use of photographic emulsions in producing and analysing electron micrographs.

A relatively inexpensive alternative to CAMERA that should become useful for studies on muscle cells is based on modifying a video camera more extensively than warranted by current commercial demands. If the associated decrease in field size and the loss in resolution are unimportant for studying the objects of interest, and if the lesser flexibility of software control over speed, format size, and synchronisation afforded by the primary sensor is not a serious obstacle, the reader may wish to explore this alternative which is described elsewhere (Roos & Parker, 1990). This latter type of system has also been added to CAMERA, where it should be particularly useful in constant monitoring of a cell, and in focussing images later captured by way of an image intensifier. The intensifier can now be gated with less likelihood that it might be damaged by an unexpectedly bright image.

All imaging with the CAMERA system is based on enhancing contrast entirely outside the optical system. The light source is imaged onto the condenser iris diaphragm, the diaphragm in front of the lamp collector lens is imaged in the plane of focus of the cell, and neither diaphragm is reduced to increase contrast. When it is necessary to capture images at rates greater than several hundred fields per second, any element that absorbs photons in the path from the back of the water immersion objective to the quartz surface on the photodiodes reduces the image intensity to such a low level that few solid-state area sensors can detect or follow changes that occur with the normal speed of events in a twitch. Hence the combination of the microscope objective, area sensor, high-speed A/D converter, image frame buffer, computer, and most importantly, the customised software for automatic acquisition and analysis, allows us to construct explicit, meaningful descriptions of physical objects from otherwise invisible images of contracting muscle. This combination has prompted workers in the field of artificial intelligence to propose a name that includes many techniques that are useful by themselves, and we have adopted this here when we discuss the field known simply as "computer vision" (Ballard & Brown, 1982).

Results and discussion

We have studied the subcellular contractile behaviour of isolated intact muscle of three types: (1) rat cardiac myocytes, (2) frog skeletal muscle fibres, and (3) newborn rat skeletal muscle grown in primary cell culture for about two weeks. The results with frog skeletal muscle that are relevant to the main points of this article have already been outlined elsewhere (Helland et al., 1988). We shall summarise some results with rat cardiac myocytes and rat myoballs in this chapter and compare them with previous results with frog muscle. The procedures for isolating and handling the cardiac cells and rat myoballs have all been described elsewhere (e.g., Roos & Brady, 1982; Boldin et al., 1987; Rolli, Wanek, & Taylor, 1988).

Cardiac myocytes. As mentioned above, shortening below a critical length causes some myofibrils to buckle and become wavy, implying that adjacent sarcomeres in skeletal muscle are physically connected to one another. The connection must be relatively noncompliant, which raised the possibility in our minds that the cytoskeleton not only accounts for this phenomenon in adult skeletal muscle, but also might account for its failure to occur in isolated cardiac myocytes (Krueger et al., 1980). Stiff mechanical links among sarcomeres may be a feature relatively specific to adult skeletal muscle (e.g., Street et al., 1966). Therefore, we examined the fine details of region-to-region shortening and relaxation in single beats of isolated rat cardiac myocytes at room temperature to test this possibility further.

The images in the top row of Fig. 9.1 are shown at three sequential stages of processing. The steps applied here were the same as the processing routine described at length elsewhere (Roos et al., 1989a). Each stage in processing was always displayed side by side with the previous stage to increase the possibility that the operator could detect the introduction of processing artefacts among the shortening and re-extending striations. After processing, a line was positioned on each segment of the optical section that appeared free of nuclei, mitochondria, or clefts. The analysis line sometimes had to be changed in length and position within the designated region to ensure that the same striations were measured. The addition or deletion of one striation from the analysis line is usually sufficient to introduce an artefactual pause in the apparent time course of shortening and re-extension. The centroid of each striation is calculated automatically, and the numerical distance between each centroid is transferred to a plotting program that prints a graph of striation spacing versus elapsed time. Hence the time course of shortening and relaxation can be measured very precisely in small, known, and distinct regions of the cell during a single beat.

Panels C–F in Fig. 9.1 show processed images of the same myocyte at different stages in a single cycle of contraction and relaxation. Plots of the time course in different regions showed that the onset of contraction was synchronous. The pattern of shortening and re-extension was uniform in a given region of a cell. However, adjacent myofibrils differed in the distance they shortened and differed also in the onset of re-extension. These findings led us to conclude that there is a degree of independence between adjacent regions of a heart cell that does not exist in a skeletal muscle cell, and that the mechanical links between fields of sarcomeres in a heart cell are relatively weak (Roos & Taylor, 1989). This means that measurements of a heart cell's overall twitch that reflect the average behaviour of its sarcomeres, such as obtained by monitoring the movement of the freely shortening ends of a cell, are subject to significant error. In addition, methods that employ coherent laser light either to monitor the striation spacing or to image striations directly may contain several sources of artefact owing to the essentially infinite depth of field (A.F. Huxley, 1986).

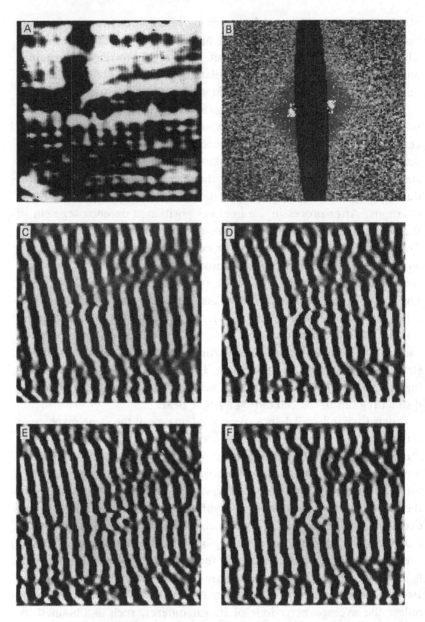

Fig. 9.1. These panels illustrate the image enhancement steps that reveal the striation pattern and contraction dynamics of a rat cardiac myocyte. (A) The results of subtracting a cell-free background are shown. The features of interest are usually concentrated in a relatively narrow range of grey scale values. This panel is frame 1 of a 39-frame sequence acquired at 170 frames per second (fps) to produce an image of a 24.85 × 24.85 μm segment near the

Skeletal myoballs. Several types of experiment are more easily done with muscle grown in culture than with adult muscle, which is well-developed and specialised for a particular contractile function. The physiology of ion channels as they develop and the chemistry of muscle-specific proteins and cytoskeletal elements are a few of the subjects in eukaryotic cell biology that are enriched with new evidence arising from the use of muscle in culture (e.g., Fischman, 1986; Baldwin & Burden, 1989; Hille, 1989; Isaacs et al., 1989; Ralston & Hall, 1989). However, far less is known about (1) the mechanisms of contraction and relaxation, (2) the mechanisms of excitation–contraction coupling, and (3) their structural correlates in such preparations (Fukuda et al., 1976).

One way to determine the function of a cell feature is to see what happens to the cell when the feature is missing. The mechanisms underlying some cell properties are often detectable only when function is disrupted by a natural or induced event. Skeletal muscle in primary culture has been a particularly useful system for electrophysiologists who have defined voltage- and ligand-gated channels (e.g., Lass & Fischbach, 1976; Horn & Brodwick, 1980; Matsuda & Stanfield, 1989). The structure, regulation, and cell type-specific expression of channels that govern electrical activity in skeletal muscle have become better known because investigators can selectively remove cell features either by the

Caption to Fig. 9.1. *(cont.)* center of a cell (R1B/88) after processing as described above. The exposure of frame 1 began 5.88 ms after the stimulus pulse. This cell had been electrically paced at 1 Hz for several beats before the sequence was taken. (B) The Fourier frequency domain of (A) with an oval mask created to remove out-of-focus features of low spatial frequency from the images, and also to thin the depth of field containing unwanted features that influence the image. The two bright spots immediately on each side of the center of the mask correspond to the dominant striation spacing. (C) In this panel, frame 1 is processed with mask B to enhance the striation pattern further. The striation spacing of the central 11 striations was measured automatically with a program that calculated the centroid of each bright band, the distance between centroids, the average band width, and standard deviations of each value; the program then submitted the numbers to a plotter that displayed each point on a graph of striation spacing versus time. The striation spacing along a line selected from frame 1 is 1.82 ± 0.06 μm in frame 1. (D) Frame 7 (at time = 41.18 ms) of this sequence was obtained when the cell was shortening at its maximum rate. The spacing of the same 11 striations measured in (C) is now 1.68 ± 0.03 μm. (E) Frame 14 (time = 82.35 ms) was obtained when the cell reached its minimum striation spacing, 1.53 ± 0.07 μm. (F) Frame 35 (time = 205.88 ms) was obtained when the cell was relaxing at its maximum rate. The striation spacing of the same 11 striations is now 1.65 ± 0.05 μm.

well-established approach of applying pharmacological agents or by deletions of selected nucleotide sequences from subunits of genes that are necessary for the muscle-specific expression of voltage- and ligand-gated channels.

Similar approaches have proven that skeletal muscle in culture is an excellent system in which to characterise physiologically important enzymes, cytoskeletal proteins, and the proteins responsible for contraction, relaxation, and their regulation. The understanding of muscle function is at a stage where the genetic expression of channels can be separated from the expression of proteins associated with the contractile apparatus (e.g., Baldwin & Burden, 1989). Myoballs have no physiological function. They have no tendons, perform no external work, and generate no force on other cells or bones. They develop without branching or with a preferred orientation and have relatively little fibrous material covering the plasma membrane. They contract and relax by a mechanism presumably identical to the sliding-filament mechanism of adult muscle. On the other hand, excitation–contraction coupling in myoballs may depend on some variation of the mechanism which seems fairly well understood in adult muscle. A large polypeptide, the ryanodine receptor, seems to be the calcium-release channel in the endoplasmic reticulum. This channel is controlled by a voltage sensor in the transverse tubular system that appears to be the dihydropyridine-sensitive subunit which allows the depolarising signal at T membranes to pass to the membranes of the endoplasmic reticulum (e.g., Rios & Pizarro, 1988). However, skeletal muscle in culture has a poorly coordinated development of its transverse tubular system and sarcoplasmic reticulum and lacks the extensive, obvious junctions between these membrane systems that are seen in adult cells (e.g., Figs. 9.2–9.5; Schiaffino et al., 1977; Kelly, 1980). This raises questions about the mechanism of stimulus–response coupling, such as, How can the events in a myoball twitch be completed in only a few milliseconds? High-speed microscopic analysis is the only direct method currently available to probe how these differences compare with what is known about contractile mechanisms and rapid signal transduction in adult skeletal muscle (Rolli et al., 1988; Morris et al., 1989). Figure 9.2 shows a low-power electron micrograph of a rat myoball about two weeks old – the age at which many features are sufficiently well developed to allow us to seek some answers. Nuclei and mitochondria are more uniformly distributed at this age, with fewer central aggregations

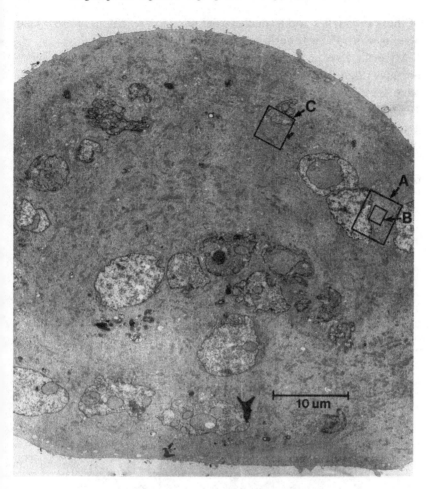

Fig. 9.2. Low-magnification transmission electron micrograph of a thin section through a 16-day-old rat skeletal muscle myoball. The portion of the myoball that rested on the bottom of the culture dish is slightly flattened. Calibration bar: 10 μm. The rectangles labelled A, B, and C contain features that are seen more easily at higher magnification. The section in this figure is about 80 nm thick. At greater magnification of the regions outlined and marked as A, B, and C, one can see that the cell is also filled with hexagonal arrays of filaments oriented at many angles. These regions were picked because they are associated with objects that are identifiable in images of the cell when it was alive, namely the surface membrane and nuclei within about 10 μm of the surface.

of these organelles and a smaller axial mass to prevent an internal load against which the myofibrils must contract.

Myoballs have a resting potential that is about 70 millivolts (mV) negative; they can generate overshooting action potentials; and they have an isopotential interior (Lass & Fischbach, 1976; Boldin et al., 1987). They also contain multiple types of voltage-gated and ligand-gated channels that are not necessarily distributed uniformly (Ruppersberg, Schure, & Rüdel, 1987; Harris, Falls, & Fischbach, 1989). Like ion channels in the myoballs, the products of nuclei that determine the proteins involved in contraction and its regulation are also probably of

Fig. 9.3. Moderately magnified transmission electron micrograph of the area outlined by rectangle A in Fig. 9.2. Corners of two of the many nuclei contained in a typical cell are the largest features. Mitochondria and myosin filaments sectioned parallel to their long axis are visible in the lower right-hand quadrant. Filaments cut predominantly in cross-section are also abundant, particularly in the area outlined by rectangle B. Calibration bar: 1 μm.

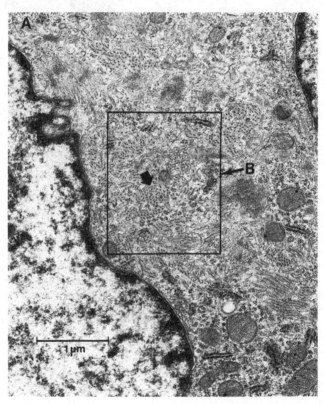

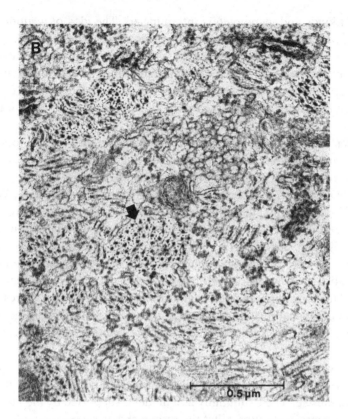

Fig. 9.4. Highly magnified transmission electron micrograph of the area out-lined by rectangle B in Fig. 9.3. Myosin filaments sectioned perpendicular to their long axis are abundant and in many places are surrounded by thin actin filaments. Neither a well-developed sarcoplasmic reticulum, transverse tubu-lar system, nor junctions among these membranes are abundant at this age, although myoballs twitch vigorously or develop strong contractures when stimulated as described in the text. When these cells were monitored with a Zeiss confocal laser scan microscope (514-nm excitation; immersion objec-tive, × 40; N.A.; 0.9), the cytoplasm filled with thin trails of fluorescent material a few minutes after lissamine rhodamine was added to the bathing solution (Endo, 1966). They remained capable of twitching for periods of illumination up to at least an hour. A network of paths in direct communica-tion with the bath evidently exists in 2-week-old myoballs just as it does in newborn rat muscle and cultured rat muscle grown on a substrate that encour-ages them to remain very flat and extensively branched (Schiaffino, Cantini, & Sartore, 1977; Kelly, 1980). There is little, if any, indication that the pre-cursor of the transverse tubular system routinely comes into close proximity with the endoplasmic reticulum. Nevertheless, a twitch often spreads through-out a myoball with a speed equal to the rate of spread in an adult frog muscle fiber at the same temperature (Gonzalez-Serratos, 1971). One possi-ble explanation for the presumed speed and relative synchrony of calcium

several kinds and incompletely intermingled. The idea of a subtle mosaic of nuclear domains in skeletal muscle cells has important implications (e.g., Hall & Ralston, 1989). However, similar to the situation with cardiac myocytes, the functional significance of differences among domains in muscle that are defined by intracellular objects can be directly confirmed or disproven only by comparing their behaviour in living cells.

When the precursor cells (myoblasts) are grown on a surface to which they cannot readily attach, the precursor cells fuse; in a few days, they form myoballs that have many nuclei and mitochondria, and they often twitch spontaneously at rates up to about five times each second. When myoballs are young, they have very few well-formed sarcomeres like those seen in Fig. 9.5, and their responses to stimulation by electrical depolarisation, depolarisation by high K^+, and mechanical response to caffeine are all very variable and unpredictable. However, when they reach about two weeks of age, they have many groups of sarcomeres aligned in series that range in length from about 3 to 12 sarcomeres long in any given section (Figs. 9.2 & 9.5). The particular myoball shown in Fig. 9.2 is 16 days old, by which age few spontaneously active cells remain.

However, almost nothing is known about the relation among activation, structure, and mechanical properties of these cells as models of skeletal muscle. Their suitability as a model is supported by the fact that electron microscopy shows that the sarcomeres in series are made up of thick and thin filaments arranged in the same hexagonal array that is typical of adult muscle (Figs. 9.3 & 9.4). In addition, although it is convenient to think of them as spheres in order to solve three-dimensional problems in electrophysiology, spherical geometry approximates the shape of most myoballs *only* when they are fully contracted. When the cells are unstimulated or when they spontaneously relax, their geometry is best approximated by treating them as spheroids that only appear to have cross-sections of constant width as they are rotated under the microscope (Blinks, 1965). Their true cross-section during contraction and relaxation is an infinite variety of closed-curved figures of changing width.

Caption to Fig. 9.4. *(cont.)* release from the endoplasmic reticulum is suggested by a point made by Dulhunty (1989). The calcium-release channels may be distributed widely over the endoplasmic reticulum rather than only at junctions with the T-system, particularly in developing muscle where the calcium-release channels may not have developed the footlike structures so prominent in the adult cell (Block et al., 1988). Calibration bar: 0.5 μm.

Using high-speed digital imaging microscopy, we do not have to make an assumption about their cross-sectional geometry or ignore the fact that it is changing during contraction and relaxation.

The implementation of the system for computer vision which we have

Fig. 9.5. Highly magnified transmission electron micrograph of the area outlined by rectangle C in Fig. 9.2. Sarcomeres with poorly developed Z-bands can be seen in series here and irregularly at other parts of a thin section. They terminate in the cytoplasm at no specific location and never appear to attach directly or indirectly to the cell surface. The number of serial sarcomeres visible in a section ranged between 3 and 12. Calibration bar: 1 μm. This is a high-power view of the third region outlined by the rectangles in the low-power view. Shown are two of the string of eight sarcomeres that are connected in series in this section. A sarcomere is bordered at each end by a dark, broad Z-band. Thick filaments are evident at the centre of each sarcomere. We have yet to see less dense regions in the middle that might be H-zones. The thin filaments are not as obvious as they are in a cross-section, and the pattern of their attachment to the large, diffuse Z-bands cannot be interpreted from these micrographs.

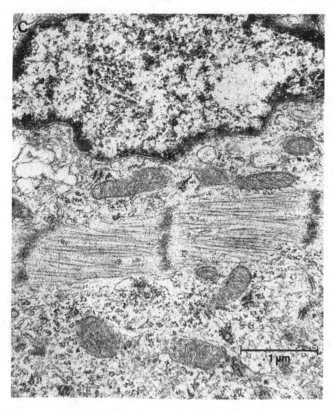

described under Methods has allowed us to study contraction and relaxation in myoballs at room temperature over a range of conditions, from conditions where they just perceptibly twitch regularly in response to an extrinsic electrical field applied at known intervals, up to the most vigourous of contractures.

Brief electrical stimuli induce shortening and spontaneous relaxation without detectable dimpling of the surface membrane. This is consistent with the fact that we have yet to see electron-dense connections between sarcomeres in series and objects at or near the surface to which they might attach. High potassium, caffeine, or carbachol cause single contractures that relax spontaneously and are not reproducible unless a myoball is rested in agonist-free solution. Solutions with no added calcium ions and EGTA (3 mM) have no effect on twitches or contractures (Rolli et al., 1988; Morris et al., 1989). Stimulus–contraction coupling in myoballs evidently depends on an easily depleted internal store of calcium ions.

In adult muscle cells it is relatively easy to locate and study the contractile machinery in the image of a living cell. However, it is more difficult in myoballs to know in what directions forces act within cells. This we can determine from digitally enhanced images (e.g., Fig. 9.6) by following the movements of regions of the cell that evidently shorten more vigorously than their neighbours. How are the force-generating elements arranged with respect to one another? How do these arrangements change during shortening and relaxation? How are these elements linked to the cytoskeleton, to each other, and to the cell surface? Little is known about the ultimate sites of myofibril termination. One possibility is they attach to intermediate filaments such as desmin; desmin is, in turn, structurally and functionally coupled to a domain that includes actin; and actin associates with adhesion sites in the plasma membrane. This model predicts that the membrane should bleb, or become scalloped or festoon during contraction (Street et al., 1966; Lazardies & Capetanaki, 1986). However, we have never seen such puckering or membrane deformability during myoball contractions, and an inextensible connection of the sarcolemma with the myofibrils is not an inextricable part of development in embryonic muscle without a long axis parallel to the substrate. The rapidly reversible blebs that may occasionally appear spontaneously in a few myoballs (Rolli et al., 1988) are not the consequence of the same insults that produce irreversible blebbing and a speedy death in other cells (Herman et al., 1988; Nicotera, Thor, & Orrenius, 1989). One possibility is that myoball blebbing may reflect a normal event, causing microtubules rapidly to form and dissolve in an oscillatory manner (Mandelkow et al., 1989).

In a myoball that obviously twitches but does so with little or no change in cross-sectional area or movement of the edges, objective measurement of motion is made easier by processing steps similar to those described above for measuring the speed and direction of movement among striations and contractile fields in cardiac myocytes (e.g., Fig. 9.6). Unlike the cardiac cells in Fig. 9.1 which have a relatively regular arrangement of contractile fields, images of the myoballs (Fig. 9.6) were in addition filtered with separable Laplacian or Gaussian masks that outlined edges of the cell membrane, organelles, and other physical objects with the same range of refractive indices. The size of a range is determined by the values used in the computer program that determine the effective thickness of an optical section after processing (Bliton et al., 1988). Both the highly organised adult muscle and disorganised myoballs shortened and spontaneously re-extended without detectable interruptions. The ends of serial sarcomeres in the myofibrils of myoballs evidently do not have a noncompliant terminus at the cell surface. The

Fig. 9.6. Steps in enhancing and outlining objects in the cytoplasm of a rat skeletal muscle myoball. (A) An image with its background subtracted and the grey scale stretched to fill most of the 12-bit range. (B) The same image processed with a Gaussian filter of variable width. The filter minimises the amplification of high-frequency noise that would normally degrade edges in the final processing step, and the variable width determines the minimum size of the features that will be outlined. The right-hand frame shows the middle image processed with a second-derivative operator (Laplacian). Real values for each zero crossing and its eight neighbours are automatically computed by the program. Then the edges are defined by interpolation to reduce their size from one pixel or greater to subpixel accuracy (Bliton et al., 1988). Objects outlined in this fashion can then be tracked with the same line measurement routine used to measure and graph the change in distance between objects with respect to time (e.g., Bliton et al., 1988). Each frame was 89.6 × 89.6 μm with the moderate degree of magnification used in these experiments (see Roos et al., 1989a).

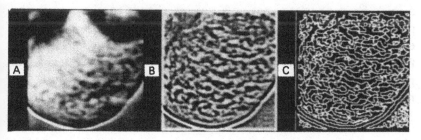

titin or connectin filaments that link thick filaments to the Z-lines at the ends of each sarcomere and reposition the thick filaments in a sarcomere after each twitch, are well developed in cultured skeletal muscle cells of the age we have studied (Horowits, Maruyama, & Podolsky, 1989; Isaacs et al., 1989). In contrast to experiments in which the reversible extensibility of connectin can be tested systematically, our studies of myoballs have shown their intrinsic ability to position thick filaments at the centre of an unrestrained sarcomere. The sarcomeres in myoballs regain their original alignment after each unarrested twitch, which supports the idea that the restoring force that re-extends sarcomeres in a shortened myofibril are intrinsic to the contractile mechanism itself (Gonzalez-Serratos, 1971).

Among the conclusions we have drawn from these studies on cardiac myocytes as well as skeletal myoballs is that edge detection is an unreliable way to monitor contraction, even in a qualitative sense, unless the contractile machinery is coupled to the edge through very noncompliant connections between the cell surface and the force generator. There was no indication that force is transmitted to the surface of a myoball in either the processed images of contraction and relaxation or the electron micrographs of myoballs. This is in contrast to the appearance of such connections in adult skeletal muscle (Street et al., 1966; Brown, Gonzalez-Serratos, & Huxley, 1984a,b).

Myoballs develop into contractile cells that have many of the phenotypic features unique to skeletal muscle. But they are not elongated tubes, they do not have tendon attachments, nor do they move significant loads over an appreciable distance. Hence they provide us with a model of functional organisation that has many qualitative features of the contractile machinery and the excitation–contraction coupling process common to adult striated muscle, as well as differences that may be significant for ascertaining what is truly general to muscle and what is relatively specific for a given type of muscle cell. For example, myoballs shorten and re-extend spontaneously very rapidly despite the absence of a thick fibrous surface coat or a volume change consistent with the possibility that restoring forces develop outside the contractile machinery. This observation is in accord with the idea that rapid relaxation is a feature intrinsic to the contractile machinery, and that its details are perhaps influenced by the characteristics of the fibrous network surrounding striated muscle of a specific type.

10

Inotropic mechanism of myocardium

LINCOLN E. FORD

Introduction

One of our main goals of my laboratory has been to learn how the heart varies the strength of its contraction in response to inotropic stimuli. Although Sir Andrew has never worked in this area directly, he has had a substantial influence here, in part because of his elucidation of contractile mechanisms and in part because of the training he has provided to workers in this field. In my own case his influence was enormous, and the experiments I shall be describing would have been impossible without the fast servo systems I learned to use in his laboratory during the three years I spent there from 1971 to 1974.

It is now well established that the contractile elements of striated muscle become activated when calcium is released into the myofilament space from the sarcoplasmic reticulum, where it is stored during diastole. One way of regulating the strength of the heartbeat, therefore, is to regulate the amount of calcium released with each beat.

A second way of regulating the strength of contraction is to alter the fraction of the released calcium that binds to the myofilaments. The binding of calcium to cardiac myofilaments is altered by well-established second-messenger systems in response to inotropic stimuli (see review by Winegrad, 1984). Taken together, the two mechanisms described so far can be classified as regulating the amount of calcium activation. A final way of regulating the strength of contraction is to alter the contractile process itself. This is the mechanism on which we have recently focussed our attention. Before we could do that, however, we had to know how changes in activation would affect the contractile properties of muscle in general and of the myocardium in particular.

Skinned skeletal muscle

Studies of intact cardiac muscle are complicated by passive viscoelastic elements both in series and in parallel with the contractile elements. Before analysing our experiments with heart muscle, we first set out to determine the effects of calcium activation on the contractile properties of a simpler system – single skinned skeletal muscle cells. This preparation does not have some of the confounding properties of intact cardiac muscle, and is therefore more likely to yield an unambiguous answer. It has no significant parallel elastic or viscoelastic elements to complicate the velocity measurements at its normal working length, and the level of activating calcium can be controlled directly and independently.

We were not the first laboratory to attempt to address this issue. Two well-respected laboratories had previously attempted to define the effects of varying calcium activation on the force–velocity properties of skinned fibres, and arrived at opposite conclusions (cf. Podolsky & Teichholz, 1970; Julian, 1971). Each laboratory produced a second, rebuttal paper defending its original position, leaving the field in a state of confusion (cf. Thames, Teichholtz, & Podolsky, 1974; Julian & Moss, 1981).

Although this seemed a dangerously controversial area, a knowledge of the effects of calcium on shortening velocity was vital for understanding inotropic mechanisms. We therefore undertook an investigation of this question ourselves. In preparation for our experiments, Dr. Richard Podolin and I wrote a fairly comprehensive review of the area (Podolin & Ford, 1983). In this review we concluded that we could not detect an obvious flaw in either set of experiments. We then began an investigation of our own, being careful to account for all of the criticisms of the previous experiments (Podolin & Ford, 1986). Our results, summarized in Fig. 10.1, shows that the force–velocity curves, obtained at an average of 45% activation, can be made to superimpose almost exactly on the curves obtained at full activation when isotonic force is nomalised to the isometric force. This is also the type of result that would be obtained if the cross-sectional area of the muscle fibre were diminished or if the number of muscle fibres pulling in parallel in a whole muscle were reduced. Thus, it appears that altering the number of cross-bridges pulling in parallel by changing activation is no different from altering the number pulling in parallel by some other means. This result implies that calcium acts simply as a switch, activating two sites at which cross-bridges attach, but not influencing the contractile kinetics of the cross-bridges once they are activated.

Our results were similar to those of Podolsky and Teichholz (1970) and Thames et al. (1974), and different from those of Julian (1971) and Julian and Moss (1981). We were not able to identify absolutely the reasons for the disparity in the results, but two possible mechanisms

Fig. 10.1. Relative force–velocity curves obtained at full (solid curve) and partial (dashed curve) activation of frog skinned muscle fibres at 3–5°C. Curves are fitted to the data of Podolin and Ford (1986). The parameters in the inset show that the force developed at partial activation is about 45% of that at full activation. Maximum velocity and relative maximum power PV_{max} are very similar, so that the two curves are almost identical. The relative maximum power is the value of maximum power (the maximum in a plot of force × velocity vs. force) normalised to the isometric force. This parameter has the dimensions of velocity and is used to indicate differences in velocity in the midrange of the curve. (Solutions contained 151 mM potassium propionate, 5 mM Na_2ATP, 6 mM $MgCl_2$, 5 mM $CaCl_2$, 4.95 mM EGTA, and 10 mM imidazole buffered to pH 7.0 at room temperature.)

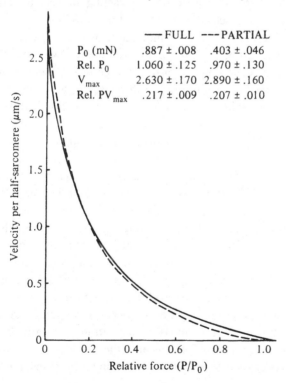

	— FULL	--- PARTIAL
P_0 (mN)	.887 ± .008	.403 ± .046
Rel. P_0	1.060 ± .125	.970 ± .130
V_{max}	2.630 ± .170	2.890 ± .160
Rel. PV_{max}	.217 ± .009	.207 ± .010

suggest themselves. First, we were not able to reproduce Julian's original conditions. He used a somewhat lower ionic strength, which caused the fibres to develop substantially more force. In our preliminary experiments, fibres usually broke when maximally activated in the lower-ionic-strength solutions, and thus we stopped trying to reproduce this condition. To be fair, we did not pursue these lower-ionic-strength conditions in later experiments, when we were more experienced and might have been successful in keeping the fibres intact. We were dissuaded from doing this by the report of Julian and Moss (1981) that the high forces generated under these conditions were associated with sarcomere disarray. Thus, one possible cause of the activation dependence of maximum velocity found by Julian might have been fibre damage, as suggested by Thames et al. (1974).

Shortening inactivation. Another possible cause of the disparity in the results is shortening inactivation. We encountered this inactivation in several forms. In our preliminary experiments, we attempted to measure the effect of activation on maximum velocity using the slack test. In these experiments, the isometrically contracting muscle is suddenly shortened so that the muscle goes slack, and the time taken for force to begin to redevelop is taken as an indication of unloaded shortening velocity. When we attempted these measurements, we encountered great difficulty in determining the onset of force redevelopment in partially activated fibres because the rate of force redevelopment was very slow. The rate of force development during full activation is rapid (Fig. 10.2G & I), as is the redevelopment of force after a sudden shortening. By contrast, the rate of force development at partial activation is very slow, both during initial activation and after a release (Fig. 10.2B & H). This slow rise of force might have been anticipated on the basis of the known cooperativity between calcium of cross-bridge binding (Bremel & Weber, 1972). Because of this cooperativity, the affinity of the myofilaments for calcium is increased when cross-bridges are bound. Activation of skinned fibres in buffered calcium solutions requires the following processes: (1) CaEGTA diffusion into the fibre, (2) dissociation of Ca from EGTA, (3) Ca binding to the myofilaments, and (4) cross-bridge attachment and force development. At supramaximal levels of activating calcium, these four processes occur in sequence and therefore fairly rapidly. At partial levels of activation, a relatively small amount of calcium is bound initially because the myofilament affinity for calcium is low. As cross-bridges attach, the affinity increases, and more calcium eventually binds to the thin filament. This binding is slowed, however, owing to both the lag caused by the finite time taken for cross-bridges to attach and the need for more CaEGTA to diffuse into the fibre and dissociate. Thus, processes 1 to 4 must occur repetitively in order for full activation to occur.

The same slow redevelopment of force is expected to occur following a quick release because rapid shortening is known to reduce the number of attached cross-bridges (Ford, Huxley, & Simmons, 1985). This reduction of cross-bridges is likely to cause a reduced calcium affinity in the

Fig. 10.2. Length and force records showing cycles of activation, isotonic shortening, quick release, and relaxation of skinned frog skeletal muscle fibres. Upper records show the standard protocol for obtaining the data shown in Figs. 10.1 and 10.3. Fibres were rapidly activated by a brief exposure to 2.5 mM free Ca solution followed by maintenance of either full or partial activation in buffered Ca solution. Partial activation (B, E, H) is always bracketed by full activation. The middle records (D–F) were made at a fast recording speed to show the time course of isotonic shortening. The curvature in the length record is greater at partial activation. The lower records (G–I) were made at a very slow recording speed, and show the very slow development of force in partial activation when there is no pretreatment with free calcium. The redevelopment of force following a sudden shortening is very slow in partial activation and much more rapid in full activation. The second shortening in these records (G–I) is used to show that the slow force recovery is not due to transducer drift.

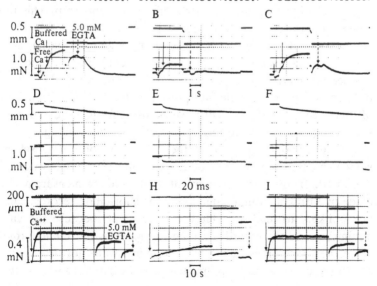

FULL ACTIVATION   PARTIAL ACTIVATION   FULL ACTIVATION

thin filaments and therefore a reduction of activation. Shortening has been shown to cause calcium to dissociate from the myofilaments, both in skeletal muscle (Ridgway & Gordon, 1984) and in cardiac muscle (Allen & Kurihara, 1982). These observations suggest that when shortening of a partially activated skinned fibre ends, the fibres must be reactivated by the same four processes described above. In practice, the development of force was so slow after a slack release that it was difficult to determine when the slack interval ended. This experience gave us a strong bias against using the slack test to determine maximum velocity in partially activated fibres.

Although our experience with slow force development would make the slack test seem less desirable as a method of measuring maximum velocity at partial activation, somewhat similar problems are encountered when the classical force–velocity measurements are used. In our experience, the length records during isotonic shortening of skinned fibres are not straight. Velocity slows continuously as shortening progresses (Fig. 10.2D–F), and this slowing is much greater in partially activated fibres. When velocity is measured immediately after the transition to an isotonic load, the resulting curves obtained at full and partial activation superimpose almost exactly (Fig. 10.3A). When the data points are measured after a substantial period of shortening, both curves are depressed, and the depression is much greater at partial activation (Fig. 10.3B). A measure of the slowing of velocity can be obtained by the ratio of the velocity measured after many milliseconds of shortening to the velocity measured shortly after the isotonic step. This ratio declines with lower loads and higher velocities at both full and partial activation, but the decline is substantially greater at partial activation (Fig. 10.3C). This finding of a curved shortening record is not new and has been observed by several investigators (Brenner, 1980, 1986; Gulati & Podolsky, 1981). Several interpretations have been offered, and none of them are mutually exclusive. Such curvature is to be expected from shortening inactivation resulting from cooperative binding.

The substantial curvature in the velocity records suggests that the velocity measurements will depend very much on the time when they are measured. In the early experiments, the velocity measurements were made by fitting a straight edge along the shortening records. This practice is likely to lead to a later measurement, because the curvature diminishes and the records become straighter as shortening progresses. Since the exact times of the measurements were not described in the earlier papers, one can only speculate on the extent to which the inacti-

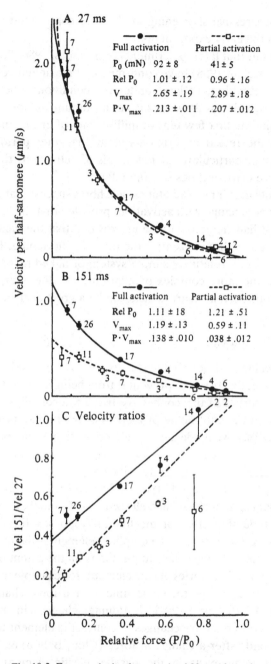

Fig. 10.3. Force–velocity data obtained 22 ms after release to an isotonic load (A) and 151 ms after the onset of shortening (B). The ratio of the late to early velocity (C) gives a measure of the slowing that occurs during shortening.

vating effects of progressive shortening might have contributed to the disparate results in the early experiments.

We should point out that later experiments (e.g., Moss, 1986; Farrow, Rossmanith, & Unsworth, 1988) have confirmed that maximum velocity is independent of activation, at least under the conditions where we measured it – that is, at levels down to 45% of maximum and for velocities measured during the first few tens of milliseconds after the onset of shortening. The same recent experiments show that after periods of prolonged shortening, particularly at low levels of activation, the unloaded shortening velocity declines substantially.

These experiments in skinned skeletal muscle fibres suggest that maximum velocity does not change with activation, provided that the level of activation is at least half the maximum and provided that the measurements are made a few milliseconds after the onset of shortening. If this conclusion, obtained in a relatively simple system of naked myofibrils, can be extended to the more complex papillary muscle, the interpretation of the results obtained in heart muscle can be a great deal simpler.

Intact cardiac muscle

Passive properties. Since the heart does not possess a skeleton, it must rely on its connective tissue to prevent it from being overstretched. The viscoelastic properties of this connective tissue can complicate substantially the study of the contractile properties of the preparation. As part of our data analysis, we made a comprehensive study of the passive properties of papillary muscle. A schematic diagram of the minimum number of passive elements needed to describe the preparation is shown in Fig. 10.4 (Chiu, Ballou, & Ford, 1982a). The approach adopted to investigate this model assumes that each element exists, and then tries to estimate the magnitude of its effect on mechanical responses. It is necessary to assume, for example, that each compliant element is damped, as shown. It then becomes relatively easy to put limits on the extent of this damping. The damping of the series elastic element, for example, causes series elastic recoil to continue for some time after a force change is complete. Because of this prolonged shortening, changes in muscle length cannot be used to measure changes in contractile element length for several milliseconds after a change in force (Chiu, Ballou, & Ford, 1982b). Similar damping exists in the parallel elastic element, and greatly complicates estimation of the load actually borne by the contractile elements.

After making a study of these passive properties, we concluded that the best approach to the study of cardiac muscle was to choose conditions where the passive mechanical element would have the least effect rather than to correct the data for the passive effects. One of the ways of minimising the complication of a parallel elastic element was to work at intermediate lengths, where the parallel spring is slack. While such short lengths minimise the effects of the parallel components, they do not eliminate them altogether. There is a substantial parallel viscoelastic element that becomes compressed as the muscle shortens. Our study of the passive properties suggested that the optimum length for measuring force–velocity properties should be between 88% and 92% of the length at which maximum force is developed (L_{max}). At that length, the force borne by the parallel elastic element is about 4% of peak twitch force. At longer lengths, the passive force is a larger fraction of the total. At shorter lengths the activated contractile elements shorten below their normal rest length, and compress an element that restores the relaxing muscle to its rest length. This compressive element creates a substantial internal load against which the contractile elements must work.

This range of lengths also corresponds to the middle of the working range of the myocardium. This is the length where we undertook our studies of activation and inotropic influence.

We also found the optimal time for making force–velocity measurements in papillary muscle at room temperature was 6–10 ms following the transition to an isotonic load. At that time the damped series elastic recoil, as well as velocity transients, have largely died away (Chiu et al.,

Fig. 10.4. Schematic representation of mechanical elements in papillary muscle (CE: contractile elements). The series (s) and parallel (p) springs exert force only when extended beyond rest length. The spring (d) in series with the parallel viscosity exerts force in both compression and extension. [Reproduced from Chiu, Ballou, & Ford (1982a). *Biophysical Journal, 40*, 109–20, by copyright permission of the Biophysical Society.]

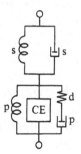

1982b). At the same time, the load on the parallel viscoelasticity has not accumulated to any great extent, and the inactivating effect of progressive shortening has not had a large effect.

Mechanical manifestation of changes in activation. We used two methods of producing changes in activation, the first being post-extrasystolic potentiation. The effect of this intervention is so rapid, within one heartbeat, that it seems unlikely to be associated with chemical changes in the contractile process. Another advantage of this intervention is that it appears to increase activation from about 40–50% of maximum up to something approaching the maximum. Thus, it appears to work over the range of activation where we had shown skeletal muscle maximum velocity does not change. We also obtained variations in activation by studying the muscles at different times during the twitch. By comparing curves obtained early in nonpotentiated contractions with those obtained at the peak of potentiated contraction, it was possible to obtain an eight-fold range of developed force.

Typical force and length records for measuring force–velocity properties are shown in Fig. 10.5. These records were reconstructed from digitised records stored in a computer. Two recording rates were used. The entire twitch was recorded at a slow rate, and the rate was increased to record the events surrounding the release to an isotonic load. Velocity was measured by fitting a least-squares linear regression to the length traces over a 4-ms interval between 6 and 10 ms after the onset of the isotonic release. The records shown in Fig. 10.5 compare control twitches with post-extrasystolic potentiated twitches. As shown, there is an approximate doubling of twitch force, but very little difference in velocity at the same relative load. Series elastic recoil is substantially larger in the potentiated contraction because the absolute force change is greater, but after this recoil the records are very nearly parallel. To gain a more exact comparison, data from multiple releases to different loads were fitted with hyperbolic functions (Fig. 10.6) and the fitted curves were compared. When the isotonic loads are normalised to developed force, the curves nearly but not exactly coincide (Fig. 10.6B). Additional curves obtained during the rise of twitch force are shown in Fig. 10.7. At the peak of the twitch (400 ms), the entire force–velocity curves can be defined by the data because velocity is very near zero when the isotonic load is equal to the developed isometric force. A different type of curve is obtained early in the twitch. A release to an isotonic load equivalent to the developed force (i.e., an afterloaded contraction) results in shortening at a substantial velocity. Force–velocity curves derived from this data can be extrapolated to zero velocity to obtain an "extrapolated isometric force" that is substantially greater than the developed isometric force. This result is very similar to that obtained in single skeletal muscle fibres (Cecchi, Colomo & Lombardi, 1978).

These results show that the force–velocity curves change both with time in the twitch and with post-extrasystolic potentiation. To describe these changes, the force–velocity data were fitted with hyperbolae (Hill, 1938) that require three parameters for description. The three parameters chosen were the two intercepts of the curve (i.e., maximum velocity

Fig. 10.5. Length and force records obtained with isotonic releases during twitch contractions of rabbit papillary muscles. The whole twitch is recorded at a slow speed, and the events surrounding the isotonic release (between the dashed lines) are recorded at a fast speed. Records from control and post-extrasystolic potentiation are superimposed. The releases are made at the peak of the twitch. Because the peak of the twitch occurred earlier in the potentiated twitch, the record has been shifted by 40 ms to the right to make the times of the releases coincide. Velocity is measured by fitting a least squares linear regression to the length record over the interval 6–8 ms following the release (between the dotted lines). [Reproduced from Y. C. Chiu, K. R. Walley, & L. E. Ford (1989). Comparison of the effects of different inotropic interventions on force, velocity and power in rabbit myocardium, *Circulation Research, 65,* 1161–71, by permission of the American Heart Association, Inc.]

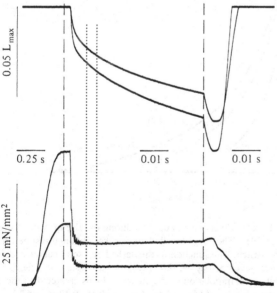

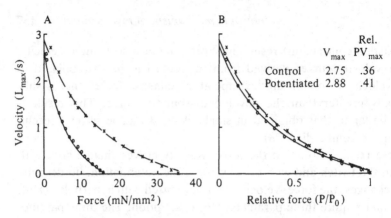

Fig. 10.6. Force–velocity curves in control and potentiated contractions. The absolute force is plotted in (A), and the force relative to the isotonic is plotted in (B). The relative curves almost, but not exactly, coincide. Maximum velocity and relative maximum power values derived from the fitted curves are shown in inset. [Reproduced from Y. C. Chiu, K. R. Walley, & L.E. Ford (1989). Comparison of the effects of different inotropic interventions on force, velocity and power in rabbit myocardium. *Circulation Research, 65,* 1161–71, by permission of the American Heart Association, Inc.]

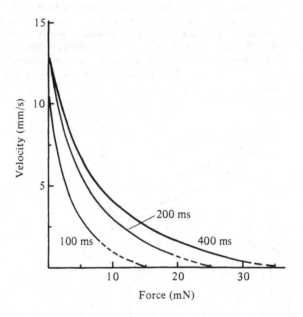

Fig. 10.7. Force–velocity curves derived during the rise of twitch force in cat papillary muscles. The curve labelled 400 ms was obtained just before the peak of the twitch. The junction of the dashed and solid curves occurs at the level of force developed at the time the data were obtained. The dashed curve is thus an extrapolation of the curves to loads greater than the developed isometric force. (Modified from Chiu, Ballou, & Ford, 1987.)

& isometric force) and maximum power. This latter value gives an estimate of changes in the midrange of the curve.

The greatest changes in the parameters of the force–velocity curves occurred in the extrapolated isometric force and the maximum power. Furthermore, these two parameters varied in almost direct proportion to each other (cf. Figs. 10.8B & C). These changes are similar to those seen

Fig. 10.8. Parameters derived from force–velocity curves fitted to data obtained during the rise of twitch force (at loads less than 100% of twitch force) and at the peak of twitch force in potentiated contractions (at loads greater than 100% of nonpotentiated twitch force). Reference values are those obtained at the peak of the nonpotentiated twitch. Cat papillary muscle. (Modified from Chiu, Ballou, & Ford, 1987.)

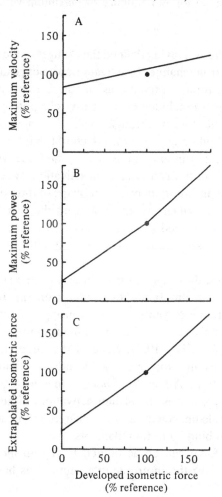

in skinned skeletal muscle fibres with changes in activation (Fig. 10.1). Maximum velocity, which has been said to signal changes in activation (Brutsaert, Claes, & Sonnenblick, 1971), changed less than two fold for an eight-fold change in developed force (Fig. 10.7A) Thus, maximum velocity did not appear to change much, although it did change a small amount, with changes of activation.

Further analysis showed that these relatively small changes in maximum velocity could be accounted for quantitatively by a small internal load, equivalent to 4–6% of the developed force in nonpotentiated contractions. Although this analysis did not in any way prove the existence of such a load, it did offer a reconciliation of the finding of no activation dependence of maximum velocity in the simpler, skinned skeletal fibre, and a small but definite activation dependence of maximum velocity in myocardium.

Inotropic interventions. Having defined the changes in the force–velocity curves that result from changes in activation, we next examined the effects of several inotropic interventions on these same curves. Among these interventions we included post-extrasystolic potentiation, in the belief that this stimulation acts through an increase in activation. We compared the effects of other agents to the effects of this intervention to see whether any of the other interventions change the force–velocity curves in a way that is different from an increase in activation. We studied three agents that have very different effects on the shape of the isometric twitch. These were caffeine, which greatly prolonged the twitch; iso-proterenol, which greatly shortened the twitch; and acetylstrophanthidin, which increased twitch amplitude without greatly altering twitch duration.

In contrast to the marked difference in their effects on the shape of the twitch, these agents had almost identical effects on the relative force–velocity curves. All four interventions caused an approximate doubling of twitch force, with only very small changes in maximum velocity and relative maximum power (Fig. 10.9). None of the three chemical interventions produced a change that was significantly different from post-extrasystolic potentiation. All of the changes that we observed could be accounted for by changes in calcium activation, either by a change in the amount of calcium released or by changes in the amount and time course of calcium binding to thin filaments.

The increase in twitch force can be accounted for by an increased calcium release in response to the inotropic intervention, as has been described by Blinks and his colleagues (Morgan & Blinks, 1982; Blinks & Endoh, 1986). This calcium release cannot explain the changes in twitch shape, however, because the twitch duration was substantially reduced in the presence of isoproterenol. This decreased duration can be explained by the additional findings of Winegrad and his colleagues (Mope, McClennan, & Winegrad, 1980) that beta-adrenergic stimulation decreases the affinity of thin filaments for calcium. This decreased affinity is probably associated with an increased rate of dissociation of calcium from the thin filaments that hastens relaxation. These two types of changes in calcium activation can explain all of the changes seen in our experiments. It was not necessary to postulate any direct effect of the intervention on the contractile elements. This does not prove that such effects do not exist, but if they do exist, they do not manifest themselves as changes in the force–velocity curves. If an inotropic intervention increases force by increasing the average force produced by an individual cross-bridge, then such an agent might be expected to cause an increase in maximum velocity. It is possible, however, that an agent might alter the cross-bridge binding by increasing the rate of attachment

Fig. 10.9. Comparison of force–velocity parameters obtained during inotropic stimulation of rabbit papillary muscle. [Reproduced from Y. C. Chiu, K. R. Walley, & L. E. Ford (1989). Comparison of the effects of different inotropic interventions on force, velocity and power in rabbit myocardium. *Circulation Research, 65,* 1161–71, by permission of the American Heart Association, Inc.]

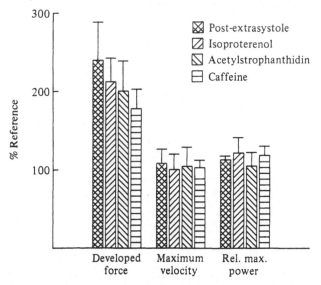

of cross-bridges to thin filaments. This type of mechanism would be indistinguishable in our experiments from an increase in activation, which also increases cross-bridge attachment rate.

Winegrad (1984) has postulated that beta-adrenergic stimulation selectively activates the fast (V_1) isoform of myosin in cardiac muscle. If our muscles contained a mixture of fast and slow (V_3) isoforms of myosin, we might have expected to see an increase in maximum velocity in the presence of isoproterenol as the V_1 myosin became selectively activated. A possible reason why we did not obtain such a result is that the adult, euthyroid animals we studied have a preponderance of the slow myosin isoform. Therefore, we are now initiating studies of the effects of beta-adrenergic stimulation on muscles having mixtures of myosin isoforms.

11

Regulation of muscle contraction: dual role of calcium and cross-bridges

ALBERT M. GORDON

Introduction

I began work with Andrew Huxley in December 1960, immediately after getting my Ph.D. in solid state physics from Cornell University and just after he had taken the chair in Physiology at University College, London. Professor Huxley, the physiologist, felt that I should be well trained in this field so I took the part 2 program in Physiology while beginning

This chapter represents the work and ideas produced in collaboration with a number of excellent colleagues. These include Ellis B. Ridgway, with whom all of the barnacle studies were performed; Lawrence Yates, who provided the initiative and expertise to develop the use of fluorescent probe-labelled TnC to investigate the calcium activation of thin filaments and much of my introduction to the biochemical literature and techniques; Taylor Allen, who collected much of the data using fluorescently labelled TnC reincorporated into the fibre, providing strong evidence of cross-bridge effects on TnC structure in skinned fibres and performed the preliminary studies on activation kinetics using "caged calcium"; and Donald Martyn, who collected much of the data on hysteresis in barnacle muscle fibres and mechanics in skinned fibres. The collaborators also include the other postdoctoral fellows in my laboratory, Bo-Sheng Pan and C. K. Wang; graduate student Frank Brozovich; medical student Bret Schulte; technicians Robin Coby, Charles Luo, and Donald Anderson; programmer and secretary Martha Mathiason; and data analyst Mark Sather. This paper is really a joint effort of all of these people, although they should not be held responsible for the manner in which the data and ideas are presented. This work has been supported by grants from the National Institutes of Health, NS-08384, AM-35597, and HL-31962, and by a grant-in-aid from the Virginia Affiliate of the American Heart Association.

my research. I am grateful for this training as I teach medical and graduate students while pursuing my research.

My initial study investigated the question raised in his paper with Rolf Niedergerke (A.F. Huxley & Niedergerke, 1958) entitled "Measurements of the striations of isolated muscle fibres with the interference microscope." Under some conditions, they had observed apparent shortening of both I- and A-bands particularly at high degrees of shortening. We reinvestigated this question using local stimulation and the interference microscope that Professor Huxley had designed and built, and found that the A-band shortening occurred only in sarcomeres that were shortening passively, being pushed by neighbouring sarcomeres that were shortening actively. In the actively shortening sarcomeres, no A-band shortening was observed, but formation of contraction bands at sarcomere lengths consistent with the sliding-filament theory was noted (A.F. Huxley & Gordon, 1962). This study showed me his dedication to finding out the truth, and not letting apparent anomalous behaviour go unexplained.

After these studies, we began to design and construct an apparatus for controlling sarcomere length to revisit the length–tension relationship. We were motivated by his observations (A.F. Huxley & Peachey, 1961) that sarcomere lengths were not uniform along the length of a single muscle fibre. Professor Huxley realised the need for sarcomere uniformity to determine the sarcomere length–tension relationship accurately and the importance of knowing this relationship precisely to test a crucial prediction of the sliding-filament theory. We built the preliminary apparatus for sarcomere length control and did the preliminary experiments. I left in September of 1962 with this work well under way and was replaced by Fred Julian, who carried on the work with Professor Huxley which culminated in the two papers on "Tension development in highly stretched vertebrate muscle fibres" (Gordon, Huxley, & Julian, 1966a) and "The variation of isometric tension with sarcomere length in vertebrate muscle fibres" (Gordon, Huxley, & Julian, 1966b).

While working with Professor Huxley, I was made aware of the possible role of calcium in the regulation of contraction by Rolf Niedergerke, who was also at University College, London, in Biophysics, by a visitor, Wilhelm Hasselbach, and by the postdoctoral fellow who joined Professor Huxley after my first year there, Saul Winegrad. Upon leaving his laboratory, I decided to investigate this question of contraction regulation which built on the muscle mechanics and optical techniques about which I had learned so much from Professor Huxley.

I shall be eternally grateful for the training and introduction to research in muscle physiology that I received from Professor Huxley. He set the highest standards for scientific excellence, experimental design, construction of sophisticated apparatus, and data analysis and theoretical calculations which have been models for me throughout my scientific life. He helped me launch a successful career in physiology and made smooth the transition from solid-state physics while encouraging my contributions from this perspective.

Introduction to the regulation of muscle contraction

Through elegant studies of Ebashi and others (see Ebashi & Endo, 1968, for a review), it has been established that calcium binding to the thin filament protein troponin activates contraction in vertebrate striated muscle. The calcium regulatory unit consists of tropomyosin, actin, and the subunits of troponin. Calcium binding to the troponin C (TnC) subunit regulates the interaction of myosin with at least the seven actins in the thin-filament regulatory unit (see the reviews of Leavis & Gergely, 1984; Ohtsuki, Maruyama, & Ebashi, 1986; and Zot & Potter, 1987, for a discussion of the interaction among the regulatory proteins).

X-ray diffraction studies of H. E. Huxley (1973) and Haselgrove (1973), and Parry and Squire (1973) led to the proposal that this regulation occurs through steric hindrance, whereby in the resting state, tropomyosin sterically hinders the interaction of myosin with actin. Calcium binding to troponin produces movement of tropomyosin such that myosin can now interact with actin (see Fig. 11.1 for a possible model of how this could happen). Studies on isolated proteins and filaments, skinned and intact muscle fibres, however, imply that the regulation of the actin–myosin interaction is more complex than this simple steric hindrance scheme would suggest. In particular, these studies point toward a more complex mechanism in which both calcium binding and cross-bridge attachment play a role in regulating actin–myosin interaction, ATPase activity, and force generation, and furthermore, that at rest the interaction is not strictly sterically hindered. The data pointing toward a more complex mechanism were reviewed some time ago by Weber and Murray (1973) and more recently by others (Eisenberg & Hill, 1985; El-Saleh, Warber, & Potter, 1986). I shall discuss some of this evidence, suggest that this combined model is the best one for explaining the control of muscle contraction, and discuss the implications of the model for contraction and relaxation.

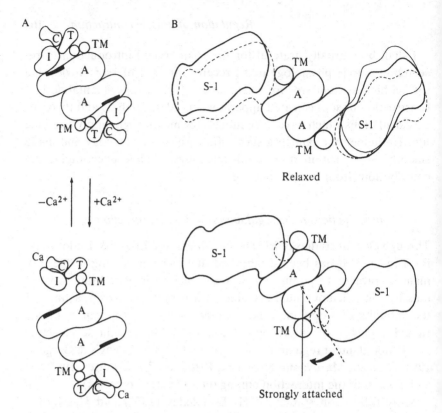

Fig. 11.1. Diagrammatic representation of calcium regulation of the thin filament of vertebrate striated muscle illustrating (A) possible changes in the
interactions of the regulatory proteins tropomyosin (TM) and troponin with its
subunits TnI (I), TnT (T), and TnC (C), and their approximate relationship
with respect to the cross-section of the thin actin (A) filament. The diagram
does not attempt to represent the precise molecular configuration. In the resting
state (panel A, *top*) both tropomyosin and TnI bind to actin [TM to seven actins
and TnI to one actin near the (dark) region where myosin can bind], stabilising
the regulatory complex on the thin filament. Calcium binding to TnC (calcium-
specific sites) increases the affinity of TnC for the region of TnI that binds to
actin, thus releasing the TnI from the actin, allowing the regulatory complex to
move to the position shown in the lower part of (A). This allows changes in the
position of tropomyosin (and possible of actin structures as well) that facilitate
the change in myosin binding status from rapidly associating–dissociating but
predominantly dissociated as in the relaxed state, to more strongly attached as
in the contracted state shown in (B). In panel (B), the cross-section of the thin
filament with interacting myosin subfragment 1 (S-1) is illustrated with the actin
(A), myosin S-1, and tropomyosin (TM) outlines as indicated in the reconstruction of Milligan and Flicker (1987). The angular change in the position of
tropomyosin is that calculated from the x-ray diffraction changes in the second
actin layer line discussed in the text. The tropomyosin position is drawn to allow
weak binding of S-1 in the relaxed state, but to retard stronger attachment
unless the TM is moved by calcium binding or the S-1 binding itself. This latter
point is discussed in this chapter.

Studies on isolated proteins and filaments

To study how calcium or cross-bridges regulate contraction, one investigates how they affect a parameter associated with contraction. In the case of isolated proteins, the parameters explored include (1) calcium or S-1 binding; (2) the local environment of one of the control proteins TnC, TnI, or tropomyosin as measured by a probe (usually fluorescent); (3) the ATPase activity; or, in the older studies, (4) superprecipitation; or (5) a combination of these. For the intact fibre this measure of activation could be force, shortening velocity, rate of force redevelopment, or structural changes inferred from x-ray diffraction. The question is, in this multistep process of activation of contraction, What measurement gives the true measure of activation? In some sense, all measure some property of activation and thus must be considered in any detailed model of regulation. This consideration must be kept in mind as we review the evidence for the dual role of calcium binding to troponin and cross-bridge attachment in activating muscle contraction.

Bremel and Weber (1972) were the first to suggest that cross-bridge attachment can modify calcium regulation of actomyosin ATPase activity by showing that rigor cross-bridges both increase calcium binding to troponin and increase the actin–myosin ATPase activity in the absence of calcium. Support for enhanced calcium binding in the presence of cross-bridges came from the studies of Grabarek et al. (1983) who showed, using fluorescent probes on TnC, that the fluorescence change accompanying calcium binding when TnC is incorporated into isolated troponin is different from the change observed when this labelled TnC is incorporated into troponin in the thin filament: cooperativity is demonstrated in the latter but not in the former. Furthermore, the cooperativity in calcium activation in the thin filament is enhanced greatly by cross-bridge attachment even in the presence of ATP. Studies by Greene and Eisenberg (1980) have shown that the binding of myosin heads [subfragment-1 (S-1) with ADP attached] to regulated thin filaments (actin, troponin, & tropomyosin) is inhibited in the absence of calcium at low S-1 · ADP concentrations, but that this binding increases with [S-1 · ADP] in a highly cooperative manner in the absence of calcium. Other studies by Trybus and Taylor (1980) and Greene (1986) have shown that attached S-1 can cause changes in the fluorescence of a probe on a subunit of troponin, troponin I (TnI), incorporated into the thin filament. These changes are similar to those produced by the addition of calcium and in some cases even greater (Greene, 1986). Furthermore, under certain conditions, the regulated actin-S-1 ATPase activity expressed per S-1 depends on the S-1 concentration even in the presence of calcium (being inhibited by low S-1 and stimulated at higher S-1 concentrations) (Lehrer & Morris, 1982). Not only are both thin filament calcium binding and actin–myosin ATPase activity of myosin heads (in the presence and absence of calcium) enhanced by attached myosin heads, but also structural changes in the regulatory proteins are produced by both calcium and bound S-1.

The biochemical studies, therefore, suggest a complex mechanism involving regulation of thin filament structure by both calcium and cross-bridge binding. The question is whether both are required for maximum activation or whether either one is sufficient by itself. Studies by Tobacman (Tobacman, 1987; Tobacman & Sawyer, 1990) show that the relationship between [Ca] and ATPase activity of regulated thin filaments still shows high cooperativity even at low S-1 concentrations. On the other hand, Greene, Eisenberg, and co-workers find that even in the presence of high calcium, both modified and unmodified S-1 further enhance the actin-S-1 ATPase activity of regulated thin filaments (Williams, Greene, & Eisenberg, 1988). Furthermore, Lehrer and co-workers find that a fluorescent probe on Cys-190 of tropomyosin that does not respond to calcium binding, changes its fluorescence on S-1 binding accompanying activation of the ATPase activity (Ishii & Lehrer, 1987). These studies in the isolated protein systems imply that, at least at high [S-1], cross-bridges are required for maximum activation measured by ATPase activity or a specific tropomyosin probe. They suggest that cross-bridge binding alone is what switches the thin filament to an activated state and that calcium binding only shifts the equilibrium toward the activated state (as in the model of Hill, Eisenberg, & Greene, 1980); however, their data do not rule out the hypothesis of dual regulation.

A mechanism of dual calcium and cross-bridge activation need not be totally different from the suggested structural changes of the steric hindrance model, which involves a shift in tropomyosin that could be accompanied by either calcium binding to TnC or myosin S-1 binding to actin (see Fig. 11.1B & Fig. 11.2, the latter of which is adapted from fig. 1 in Hill, Eisenberg, & Greene, 1983). In both cases the result is apparently the modification of both the tropomyosin position and the actin–myosin interaction to activate the ATPase. The "fully activated state" brought on by rigor cross-bridges (Williams, Greene, & Eisenberg, 1988) is another issue, and one must investigate whether this occurs in the more ordered system of the muscle.

Studies on skinned fibres

Because in the biochemical studies the S-1 concentration is an important parameter in the regulation by S-1, one must test any model of regulation in the ordered filament lattice system where S-1's are in relatively fixed positions. The skinned fibres offers such a system. Furthermore, in skinned fibres, one can also measure the mechanical parameters of muscle force, shortening velocity, stiffness, and rate of force redevelopment as well as ATPase activity, calcium binding, and the local environment of some of the regulatory proteins, using fluorescent probes. This allows one to investigate in this system many of the steps in the regulation of muscle contraction.

Studies from skinned fibres support dual regulation by calcium and cross-bridges. Rigor cross-bridges can activate force generation in skinned fibres in the absence of calcium, as was shown some time ago by Reuben et al. (1971) and more recently by Goldman, Hibberd, and Trentharm (1984b), using caged ATP. Rigor (Fuchs, 1977) and force-

Fig. 11.2. Model of activation of the thin filament by both calcium and cross-bridges. In the thin filament, tropomyosin (Tm) binds to seven actins and overlaps the neighbouring tropomyosins. Calcium binding to troponin (Tn) (top row) causes movement of tropomyosin which is coupled to the neighbouring tropomyosins, partially moving them also. This allows for the stronger attachment of myosin to a number of actins. Attachment of myosin in a strongly bound configuration (left) also moves tropomyosin, affects neighbouring tropomyosins, and affects troponin, thus increasing calcium binding. In this diagram, tropomyosin movement represents thin filament activation and increases with both calcium and cross-bridge binding, although this has not been proven for the intact fibre (see text). [Reproduced by permission from L. D. Yates (1985). *Reciprocal Meat Conference Proceedings, 38*, 9–25. Adapted from Hill, Eisenberg, & Greene, 1983.]

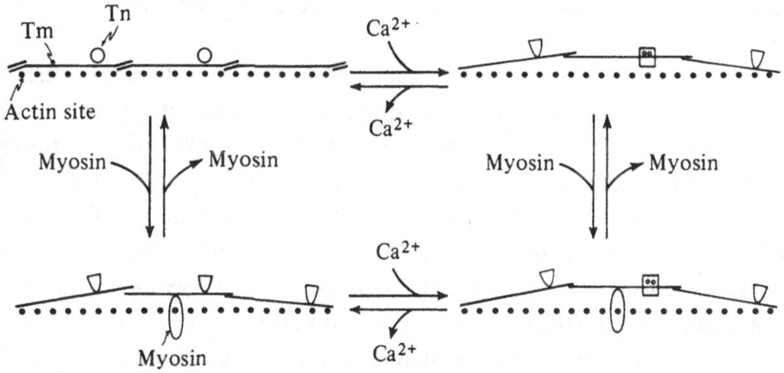

generating cross-bridges (Hofmann & Fuchs, 1987) also enhance calcium binding and affect the TnC structure as measured by a fluorescent probe on TnC in much the same way as does calcium binding (Guth & Potter, 1987; Gordon et al., 1988). Furthmore, active cycling cross-bridges produce the largest change in fluorescence of reincorporated TnC labelled with a dansylaziridine probe near the calcium-specific sites, suggesting that cycling cross-bridges may be more effective at modifying the TnC structure (activation) than are rigor cross-bridges (Guth & Potter, 1987; Gordon et al., 1988).

The force–calcium relationship in skinned fibres suggests a more complex calcium activation scheme. The steady-state force in skinned fibres is a very steep function of calcium concentration, showing high co-operativity, and depends on the [MgATP], [Mg], [PO₄], and sarcomere length and ionic strength (see Yates, 1985). If one interprets this relationship as indicative of calcium binding, it is steeper than calcium binding to the thin filaments alone in the absence of cross-bridge interaction (see Fig. 11.3; Allen & Gordon, 1989). Brandt et al. (1987) suggest that this steepness implies activation of the entire thin filament as a unit. However, studies of Moss, Giulian, and Greaser (1986) extracting whole troponin to produce non–calcium-sensitive force suggest that this may not be the case. In contrast, Brenner (1988a) suggests that calcium, rather than binding to TnC to modify the number of actin sites to which cross-bridges can attach, instead modifies the rate constant for cross-bridge attachment. With this type of control a steep dependence of force on calcium can arise from a much less steep dependence of the attachment rate constant on calcium. In either case, the steady-state studies of calcium activation indicate that both calcium binding and cross-bridge attachment can activate force production and that the calcium dependence of force production is steeper than expected from calcium binding alone.

Studies in skinned fibres suggest that calcium and cross-bridges rather than controlling attachment of cross-bridges in an absolute, steric hindrance, manner, control the transition between a weakly bound, rapidly associating/dissociating state to a more strongly attached state (as shown in Fig. 11.1B). Brenner et al. (1982) found evidence for this weakly attached state in resting fibres without calcium by stretching the fibre rapidly enough, faster than the dissociation/reassociation rates. This rate-dependent stiffness was also a function of ionic strength – increasing at low strength but still measurable at more physiological ionic strengths. Because active contractions involve more strongly at-

tached cross-bridges, this result suggested that activation involves a
regulation of the transition from weak to strong attachment. The data
in skinned fibres support the idea that both calcium and cross-bridges
enhance this transition.

Analysis of calcium activation transients can provide further insight
into the possible dual activation mechanism. Recent studies in skinned
muscle fibres, using caged calcium and a fluorescent probe on TnC to
monitor thin filament structural changes, examining muscle stiffness to
monitor cross-bridge attachment, and using force to measure contraction, suggest that for a rapid stepwise increase in free calcium to reach
levels above that required for maximal calcium activation, the structural

Fig. 11.3. Steady-state force–calcium relationship in a skinned rabbit psoas
muscle fibre in the presence of 130 mM K, 3 mM free Mg, 3 mM MgATP, 15
mM creatine phosphate, 10 units/ml of creatine phosphokinase (CPK), pH
7.0, with at least 20 mM pH buffer (MOPS), keeping ionic strength at 0.2 M,
15 mM EGTA with calcium added to adjust the pCa ($-\log_{10}$ [Ca]) to the
indicated value. The data points for the active force (normalised to unit
cross-sectional area using the measured fibre diameter) versus pCa are fitted
using a nonlinear least square fit to the Hill equation: force/force$_{max}$ = 1 +
$10^{n(pCa-pCa_{1/2})}$). The Hill parameters fitting this data are pCa$_{1/2}$ = 5.61 and n =
3.9. Note how steep this relationship is. A *simple* calcium binding curve
would go from 10% to 90% over a factor of 81 in calcium concentration.
This curve does so in a factor of about 3. The apparent cooperativity in this
curve is discussed in the text. (Data from an experiment of Dr. L. D. Yates,
used with his permission.)

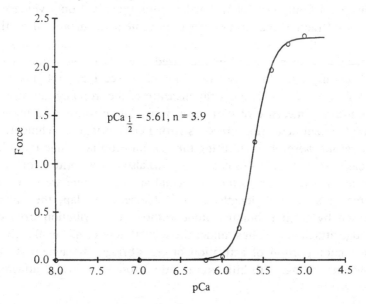

changes in TnC associated with calcium binding must precede the increase in muscle stiffness (strong cross-bridge attachment) and force (T. Allen et al., 1989). The fluorescence changes associated with cross-bridge attachment occur later and are much smaller in amplitude. This suggests that if the fluorescent probe on TnC reports the state of activation of the thin filament, then for maximal, rapid calcium activation, cross-bridge attachment may have only a small role in the onset of contraction as it is slower than the calcium-binding-induced activation. However, these data do not apply to lower levels of calcium activation where cross-bridge attachment occurring later may play a synergistic role with calcium binding to increase activation either by a direct cross-bridge effect or by a cross-bridge-induced increase in calcium binding. I shall return to this point later.

Studies on intact muscle

Intact muscle is the ultimate test of a dual model of activation. In intact muscle, one can measure all the mechanical parameters to assess activation of contraction, and using x-ray diffraction techniques, one can also investigate the structures involved in regulation and cross-bridge attachment.

In intact muscle, muscle stiffness (measured by rapid length changes) precedes force development (Cecchi, Griffiths, & Taylor 1986; Ford, Huxley, & Simmons, 1986). Furthermore, there is strong evidence from x-ray diffraction that tropomyosin movement (as measured by the increase in the second actin layer line intensity) precedes cross-bridge attachment (as measured by increased intensity of the equatorial [1,1] reflections) after the onset of stimulation (see Fig. 11.4, taken from Kress et al., 1986). If the x-ray measure of the tropomyosin position is an accurate measure of thin filament activation, these data imply that thin filament activation precedes strong cross-bridge attachment. Similar results were obtained using the 5.9-nm actin layer line intensity increase except that in this case there was also a cross-bridge component. Since the x-ray changes for the second actin layer line are virtually the same at sarcomere lengths beyond filament overlap, they must be caused by calcium binding alone without a contribution from cross-bridge attachment. Thus, under these conditions of maximal activation, the major portion of activation of cross-bridge attachment is due to calcium binding, with little contribution from cross-bridge attachment.

The question is, What happens at submaximal activation? Definitive studies on this point remain to be done.

These same studies of Kress et al. (1986) also demonstrate a role for cross-bridge attachment in prolonging activation and contraction. Kress et al. (1986) (see Fig. 11.5) show that when the muscle is stretched to beyond thick and thin filament overlap, the second actin layer line intensity declines much more rapidly after electrical stimulation ceases than when the muscle is at a sarcomere length of maximum filament overlap. In the latter case, the second actin layer line amplitude declines with about the same half-time as the muscle tension, implying that the tropomyosin returns to its resting position only as cross-bridges detach. Insofar as the increased intensity of the second actin layer line indicates tropomyosin movement and is a measure of thin filament activation, then this activation is influenced strongly by cross-bridge attachment.

Fig. 11.4. Time course of the sequential changes in the x-ray diffraction patterns of muscle compared with the rise in tension after the first stimulus. Plotted as a function of time after this first stimulus are the second actin layer line intensity (filled squares; indicated from modelling studies of tropomyosin position), equatorial [1,1] reflection intensity increase (filled circles; indicative of myosin S-1 movement toward actin), and tension (open circles), all in the same muscle at 5°C. Note that the second actin layer line leads the equatorial changes which in turn lead the tension rise, consistent with the idea that calcium-induced tropomyosin movement precedes strong cross-bridge attachment which in turn precedes tension generation. [Reproduced by permission from M. Kress et al. (1986). *Journal of Molecular Biology, 188,* 325–42.]

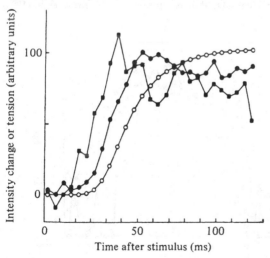

These studies, however, cannot differentiate between (1) an effect of cross-bridge attachment to prolong calcium binding and thus delay the return of tropomyosin during relaxation to the resting position, and (2) a direct effect of cross-bridges on tropomyosin position, leading to thin filament activation, although these effects should be coupled.

In intact fibres the steady-state relationship between calcium and force is very steep, even steeper than that seen in many skinned muscle fibres (Yue, Marban, & Wier, 1986) with Hill coefficients of about 5. In

Fig. 11.5. Time course of the changes in the second actin layer line (squares) and tension (circles) during a tetanus. The amplitude (square root of the intensity) of the second actin layer line is plotted against the time after the first stimulus for two sarcomere lengths – rest length (solid squares) and length when stretched to beyond filament overlap (open squares). The arrows at the top indicate the times of the stimuli delivered at 20 Hz. The time scale has been changed during the trace from 10 ms per frame to 50 ms per frame after 290 ms to show both the rising and falling phases at the appropriate rate, in order to demonstrate the relative time courses. Note that during the rise of activation, the amplitude of the second actin layer line rises over about the same time course with and without filament overlap preceding tension development. During relaxation, there is a large difference in the time course of the second actin layer line amplitude with filament overlap: the amplitude declines early with no filament overlap, and later at nearly the same time as tension with filament overlap. Insofar as the amplitude of the second actin layer line reflects movement of tropomyosin, this implies that tropomyosin movement during activation occurs early and does not depend on cross-bridge attachment whereas when there are cross-bridges attached during the plateau of tension, they retard the return of tropomyosin to its rest position. [Reproduced by permission from M. Kress, et al. (1986). *Journal of Molecular Biology, 188,* 325–42.]

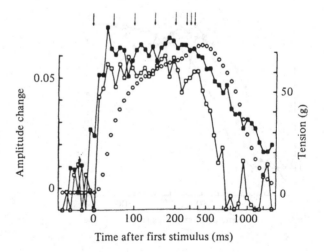

these studies on intact tetanised cardiac muscle, intracellular calcium was varied by changing extracellular calcium and measured by following aequorin luminescence. Although many assumptions are made in these calculations, the data are supportive of steep calcium activation, showing possible high cooperativity consistent with a dual role of calcium binding and cross-bridges in activation.

Evidence that cross-bridge attachment influences calcium binding comes from the studies of the effect of muscle length changes on the free calcium in muscle fibres during the muscle calcium transient measured with aequorin (Gordon & Ridgway, 1978; Allen & Kurihara, 1982; Housmans, Lee, & Blinks, 1983; Ridgway & Gordon, 1984). In these studies, a length decrease during the declining phase of the calcium transient (a decrease large enough to detach cross-bridges) leads to a drop in muscle force and a rise in *free* calcium (*extra* calcium) above the control calcium transient (see Fig. 11.6). We (Ridgway & Gordon, 1984) and others have demonstrated that this extra calcium derives almost certainly from the activating sites (presumably on the thin filaments since all of these muscles are thin filament–TnC regulated). The extra calcium has charac-

Fig. 11.6. The effect of extra light (i.e., calcium) on both shortening and the biphasic response to stretch of a barnacle single muscle fibre, voltage-clamped, length-controlled, and microinjected with the calcium-luminescent photoprotein aequorin. Muscle length, membrane voltage, muscle force, and light (luminescence from the injected aequorin as it binds calcium) were recorded as the muscle fibre was either shortened (A) or stretched (B) by 3% of the initial muscle length. The extra light in response to the shortening step or stretch was computed by subtracting a control light record with no length change from the experimental records. Note the effect of extra light (implying extra calcium released from the filaments) upon fibre shortening: a large decrease in force and cross-bridge attachment is produced. Note in contrast the biphasic response to stretch: a large increase in calcium accompanies the large stretch which presumably detaches cross-bridges, and a decline in calcium occurs subsequently as calcium rebinds to some sites (possibly the strained thin filaments) accompanying the elevated force after stretch. (Data collected with Ellis B. Ridgway, such as in Gordon & Ridgway, 1990.)

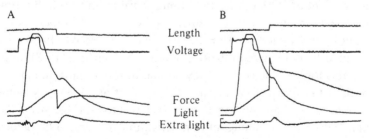

A   B

Length
Voltage

Force
Light
Extra light

teristics that one would expect from filament–bound calcium: (1) for the shortening steps at different times during the declining phase of the calcium transient, the peak amplitude of the extra calcium has a time course that is intermediate between the free calcium and force; and the extra calcium increases either (2) as the muscle force is increased by increased stimulation or (3) during force redevelopment after a shortening step (Ridgway & Gordon, 1984; Gordon & Ridgway, 1987). We (Ridgway & Gordon, 1984) and others (Allen & Kurihara, 1982) have hypothesised that this extra calcium derives from the activating sites on the thin filament and is released owing to a decrease in calcium affinity of these TnC sites accompanying the cross-bridge detachment brought on by the decreased muscle length.

Additional studies (Gordon & Ridgway, 1990) of the effects of length changes in the aequorin-injected barnacle muscle fibre demonstrate that if the muscle length is increased rather than decreased, there is a biphasic change in sarcoplasmic free calcium, increasing during the muscle stretch, and decreasing to below control while the muscle is maintained in the stretched, enhanced force state (see Fig. 11.6B). The biphasic response can be explained by (1) cross-bridge detachment that must occur during the large-amplitude stretch, causing decreased calcium binding; and (2) strained cross-bridges that occur after the stretch, bringing about enhanced calcium binding to above the control level. These data are consistent with an effect of both attached cross-bridges to increase calcium binding to the thin filament activation sites and strained cross-bridges to further increase calcium binding.

Intact contracting muscle fibres can show an effect of length change in which the muscle fibre produces either less or more steady force at the new length than expected from the isometric force–length relationship. Edman (1975) showed that during nonmaximal contractions in intact frog muscle fibres, shortening steps produced a deactivation, a decreased force at that new length, which lasted several seconds but which in skinned fibres was most prominent for partial activation. Stretch enhancement of force (above the isometric value) in tetanically stimulated muscle fibres has been described by Abbott and Aubert (1951) and extensively studied by Edman, Elzinga, and Noble (1978), Colomo, Lombardi, and Piazzesi (1989), and Sugi and Tsuchiya (1988). In particular, this force enhancement may be due to rapid reattachment of cross-bridges that have been strained (Colomo et al., 1989b). Whether this enhanced calcium binding is an effect of cross-bridge strain on thin-filament calcium binding and on activation remains to be determined.

That force deactivation with shortening and force enhancement with stretch are due to effects of cross-bridge strain on thin filament activation is by no means proven, but in light of the apparent effects of cross-bridge strain on calcium binding in the barnacle muscle fibres discussed above, this point needs to be further investigated.

Conclusions and implications

Taken together then, there is strong evidence from studies on isolated muscle proteins, skinned fibres, and intact muscle fibres that both calcium binding and cross-bridge attachment play roles in the regulation of contraction and that a more complex model of regulation is consistent with the data. The questions are (1) whether both are physiologically important mechanisms of force regulation, (2) whether there are conditions in muscle where either mechanism is more important, and (3) whether one can differentiate between their effects in the activation of muscle. The x-ray data (Kress et al., 1986; Figs. 11.4 & 11.5) and the caged calcium data (T. Allen et al., 1989) under maximal calcium activation both suggest that calcium binding alone and the subsequent thin filament changes (tropomyosin movement and changes in actin as measured by x-ray diffraction and TnC environment as measured by TnC label fluorescence) are sufficient to bring about the rise in stiffness indicating cross-bridge attachment. However, the eventual rise in force is delayed from the rise in stiffness. Thus, since cross-bridge attachment follows the indications of calcium activation, the most logical explanation is that it cannot be responsible for activation. Furthermore, both the x-ray diffraction and TnC fluorescent-probe indications of thin-filament structural changes occur at sarcomere lengths beyond filament overlap; thus these changes cannot involve cross-bridge attachment.

In addition to implications about the initial activation, both studies cited above show that these measures of thin-filament activation do not change much more with cross-bridge attachment; however, they do not rule out a cross-bridge effect that would produce no significant changes in either tropomyosin position or TnC structure. In fact Bagni, Cecchi, and Schoenberg (1988) propose that the lag between force and stiffness is due to a cooperative transition of cross-bridges between a low- and high-force-producing state, dependent on the number of attached cross-bridges. This model would provide a role for attached cross-bridges in the activation of force.

The x-ray diffraction studies of Kress et al. (1986) and the caged calcium studies of T. Allen et al. (1989) also suggest that the calcium binding step is rapid, occurs first, and is responsible for triggering the rapid cross-bridge attachment that produces high stiffness and an intensified [1,1] reflection. These data are also supportive of the role of cross-bridge attachment in prolonging muscle contraction and steepening the relationship between calcium and force in the steady state by increasing the slowly developing and steady-state response at low calcium levels. The barnacle data showing decreased calcium binding with cross-bridge detachment and increased calcium binding with cross-bridge attachment enhanced by cross-bridge strain further support a role for cross-bridge attachment in regulation. This role is to enhance calcium binding, but the activation of ATPase and of force by rigor bonds (Bremel & Weber, 1972) and the enhancement of the fluorescence of the TnC probes by cycling cross-bridges imply a more direct effect of cross-bridges on contractile activation. Thus the dual calcium–cross-bridge regulation of contraction – by (1) calcium binding to the thin filament and (2) cross-bridge attachment to the thin filament – is implied by the data obtained from contractile proteins, skinned fibres, and intact muscle fibres.

The above discussion does not try to answer the question of which step or steps in the cross-bridge cycle is or are regulated by calcium. The x-ray data of H. E. Huxley (Kress et al., 1986) support strong attachment of cross-bridges as the regulated step. The biochemical data of Greene and Eisenberg (1980) and the skinned fibre data of Brenner (1986) support the idea that calcium regulates a rate constant leading to strong cross-bridge and force production. Calcium does not regulate the rate-limiting step in phosphate release (as originally proposed by Greene & Eisenberg, 1980), which is an isomerisation that is insensitive to calcium, but does regulate an earlier step in the kinetics which could be involved in a weak-to-strong attachment of an $AM \cdot ADP \cdot P_i$ state. With the realisation that weak cross-bridge attachment does occur in resting fibres in the absence of calcium, it would seem that most data now support the view that calcium controls strong binding of actin and myosin (with ADP and P_i attached) as diagrammed in Fig. 11.1B.

There are a number of implications for the regulation of muscle contraction by a dual calcium–cross-bridge system. This system allows striated muscle fibres to be "turned on" rapidly, in accord with their roles in rapid body movements. Speed is achieved because the free calcium is elevated rapidly, and the calcium binding and subsequent tropomyosin movements also appear to be very rapid as shown by the x-ray data (Kress et al., 1986). However, this large elevation in free calcium required to "turn on" the system quickly is not required to sustain contraction. The activating effect of cross-bridges keeps the thin filament activated by both increasing the calcium affinity of the TnC sites and acting directly on the thin filament. With less calcium required, contraction can be maintained more efficiently since less energy will be expended by the sarcoplasmic reticulum in transporting calcium. Contraction is maintained more efficiently and longer than one would predict purely on the basis of electrical activation of the striated muscle fibres.

The prolongation of contraction beyond the time of electrical activation is a well-studied phenomenon in neurophysiology. Partridge (1966) and others have described the hysteresis observed in force as the frequency of stimulation of the motor nerve to a muscle is increased and decreased. The phenomenon has also been described at the cellular level by Blaschko, Cattell, and Kahn (1931) (as a long-lasting enhancement of force with a single extra stimulus inserted in a train of stimuli to a single muscle fibre) and by Wilson, Smith, and Dempster (1970) (as the hysteresis in force as the frequency of nerve stimulation is increased and decreased or the enhanced force to a given membrane depolarisation if the fibre has just experienced a greater depolarisation). We have shown in barnacle muscle fibres that there is a hysteresis in calcium sensitivity such that the steady force for a given calcium concentration is greater if the stimulation protocol is such that the free calcium is stepped down to that value rather than stepped up to that value. This was shown in both skinned and intact, aequorin-injected muscle fibres (Ridgway, Gordon, & Martyn, 1983). Thus there is strong support from a number of experiments that once "turned on," the system tends to remain activated and that this persistent activation could well be due to the attachment of cross-bridges. All these effects are expected from a dual calcium–cross-bridge regulation scheme.

12

Fibre types in *Xenopus* muscle and their functional properties

JAN LÄNNERGREN

Introduction

I came to work at University College, London, in the autumn of 1973. Andrew Huxley – "Prof" as I soon learned to call him – and I had originally discussed doing tension–transient experiments on slow-tonic fibres, which I had worked on previously. It turned out that the equipment was heavily occupied by other experiments during that autumn, so we decided instead to follow up an old observation by Caspersson and Thorell (1942) that adenine nucleotides appear to be compartmentalised in muscle fibres and might change their distribution upon stimulation. The idea was typical of Prof's sharp eye for older, potentially important observations, and we started to design a setup for the experiments with high hopes. I learned an invaluable amount of basic optics and microscopy during this period, and also that fibres are very easily damaged by intense, short-wavelength, ultraviolet (UV) light. In the end, the results were disappointing: we found no clear evidence of compartmentalisation and certainly no stimulus-related changes in the UV absorbance pattern (Lännergren, 1977).

The negative result meant that there was little incentive to pursue this line. Thus, back in Stockholm I continued my work on contractile properties of various fibre types in *Xenopus*, now armed with a much better understanding of cross-bridge kinetics after many illuminating sessions with Prof, Bob Simmons, Lincoln Ford, and Yale Goldman. The realisation that there is a broad range of shortening velocities among *Xenopus* fibres then led us to attempt to sort out the various isomyosins in this species, the results of which are summarised below.

The period 1973–4 was an extremely exciting one in Prof's lab, because it was then that the majority of the new classical tension–transient experiments were done. I often look back to that time, remembering it as one of the most stimulating periods of my scientific life.

Muscle fibre types

A. F. Huxley, while pursuing his fundamental studies on the role of the T-tubule system in activation and on the cross-bridge mechanism, was also interested in the diversity of muscle. Thus at one stage, in collaboration with L. D. Peachey, he studied slow-tonic fibres and their activation mechanism. Slow-tonic fibres actually fall at one end of a spectrum of fibre types in amphibian muscle, whereas fast-twitch "white" fibres fall at the other end. I have done some research on those two types and also on intermediate fibre types. Some of that and related work are summarised here; in the later part of this chapter, recent results bearing on the problem of muscle fatigue are also described.

A muscle fibre can be characterised by two main properties: speed and endurance. Variability in these two aspects will be discussed in turn. Speed may relate either to rapidity of the twitch or to isotonic shortening velocity. The former is influenced by the efficacy of the activating system, whereas the latter reflects the turnover rate of cross-bridges. Turnover rate is a property of the myosin isozyme in the fibre; thus differences in shortening velocity indicate the presence of different myosin forms.

Isomyosins in Xenopus *muscle*

Initial studies indicated the presence of five major fibre types in *Xenopus* skeletal muscle (Lännergren & Smith, 1966), and later investigations demonstrated distinct force–velocity relations for these types (Lännergren, 1978, 1979; Lännergren, Lindblom, & Johansson, 1982). Figure 12.1 shows in schematic form the distribution of the five types in the iliofibularis muscle and the ranges of their maximum shortening velocities at 20–23°C.

In two studies (Lännergren & Hoh, 1984; Lännergren, 1987), myosin analysis was performed on fibres of various kinds, dissected from the iliofibularis muscle. Extracts of single fibres, previously characterized by their force–velocity relations, were applied to pyrophosphate gels and subjected to electrophoresis under nondenaturing conditions. The upper part of Fig. 12.2 summarises the results. Each major fibre type (denoted by a filled circle) yielded a specific pattern, with three bands for the two fastest types (1, 2) and a single band for types 3–5. Sodium dodecyl sulfate (SDS) gels, displaying heavy (HC) and light (LC) chains of myosin, were also run, and the interpretation of the collected results is shown in the lower part of the figure. Although not clearly indicated in the figure, the HCs of the five types are likely to be different. The LCs of types 1 and 2 are the same (LC1f, LC2f, LC3f), and different combinations with the HC give rise to three isomyosins for types 1 and 2, respectively. Type 3 lacks LC3, and its other light chains (LC1s and LC2s) migrate more slowly than their counterparts in types 1 and 2. LC3 is also missing in types 4 (intermediate) and 5 (slow-tonic) fibres. The notion that the HCs are type-specific for all five types receives support from recent reports on newly available antibodies that recognize the five types individually (Rowlerson & Spurway, 1988; Bennett, Davies, & Everett, 1989).

Myofibrillar ATPase activity

Another way of elucidating myosin differences is to investigate myofibrillar ATPase activity. In our first study, Smith and I tried conventional (mammalian) ATPase stains with ambiguous results, and later Smith and Ovalle (1973) denoted as "dark" the myosin ATPase staining of types 1,

Fig. 12.1. Schematic drawing of distribution of major fibre types in the iliofibularis muscle of *Xenopus*. V_{max} values for types 3–5 are from previous publications; V_{max} ranges for types 1 and 2 include more recent data obtained with ramp releases and slack tests.

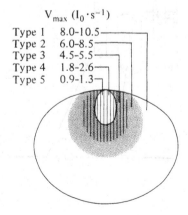

$$V_{max} \ (l_0 \cdot s^{-1})$$

Type 1 8.0–10.5
Type 2 6.0–8.5
Type 3 4.5–5.5
Type 4 1.8–2.6
Type 5 0.9–1.3

2, and 3, which differ in contractile speed. More recent investigations, however, demonstrate a rank order of ATPase activity for all five types, type 1 being highest and type 5 the lowest (van der Laarse, Diegenbach, & Hemminga, 1986; Rowlerson & Spurway, 1988). In quantitative histochemical studies, with various ATP concentrations, van der Laarse et al. demonstrated that the K_m for the ATPase reaction is higher in type 1 fibres than in the other types; they also showed a good correlation between the maximum myofibrillar ATPase activity and published values for maximum shortening velocity of types 1–5.

A different way of looking at myosin ATPase activity involves the use of heat measurements. Heat production can be measured in single fibres, as first demonstrated by Dr. R. C. Woledge and his collaborators (Curtin, Howarth, & Woledge, 1983). This technique has been applied to single *Xenopus* fibres of types 1–4 by Elzinga, Stienen, and myself (Elzinga, Lännergren, & Stienen, 1987), using metal–film thermopiles. Force–velocity properties were first measured, and the fibre was then transferred to the thermopile, where measurements were made during isometric contractions. Successful measurements were obtained for a total of 22 fibres; the difference in the stable maintenance heat rate (h_b)

Fig. 12.2. *Top:* Schematic representation of isomyosin distributions, as seen on pyrophosphate gels, for different fast-twitch varieties (1n to 2n), transitional forms (2n/3), and slow-twitch (3), intermediate (4), and slow-tonic fibres (5). *Bottom:* the distribution of light chains (LC1 and LC2) as deduced from SDS gels is shown. Only major fibre types (denoted by filled circles) are illustrated. LC2 is omitted for clarity; it is more slowly migrating in types 3–5 (LC2s) than in types 1 and 2. Types 1s and 2f are varieties of the basic types 1n and 2n; 2n/3 is a transitional form between types 2 and 3. (Modified from Lännergren, 1987.)

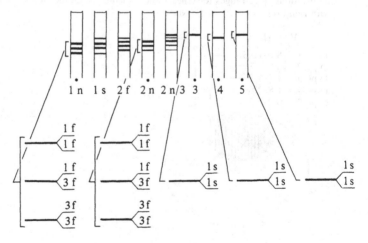

between the fastest fibres (type 1) and the single type 4 fibre was about 15 fold. With values from the literature for myosin subunit-1 (S-1) concentration, the free enthalpy change for phosphocreatine splitting, and the proportion of activation heat, we used the h_b values to calculate cross-bridge turnover rates; these turned out to be 11.6, 7.0, and 2.9 s^{-1} for types 1, 2, and 3, respectively (at 20°C).

ATP production capacity

Speed, as mentioned, is one parameter of muscle performance; endurance, or resistance to fatigue, is another. Obviously, endurance is likely to depend on two major factors: (1) the rate of energy expenditure, and (2) the rate of energy supply. Energy can be liberated either anaerobically or aerobically, the latter being more advantageous. The early studies (Lännergren & Smith, 1966; Smith & Ovalle, 1973) noted large differences in mitochondrial density among fibres in the iliofibularis muscle. More direct evaluations of the oxidative capacity have since been made, by measuring the oxygen consumption of single fibres. The maximum rate of O_2 consumption (Vo_2; recorded during twitch trains of increasing frequency) was about 5.5 times higher in type 3 than in type 1 fibres, with type 2 fibres intermediate between them (Elzinga & van der Laarse, 1988). A strong correlation between histochemically determined succinate dehydrogenase (SDH) activity and in vivo maximum Vo_2 measurements was obtained. The contribution by anaerobic glycolysis to ATP production of fibres is not known but may be higher in type 3, because in amphibian muscle, in contrast to mammalian muscle, there is a direct correlation between oxidative and glycolytic capacity (Spurway, 1985).

Thus, for *Xenopus* twitch fibres, the following rank orders apply:

ATP production rate: type 3 > type 2 > type 1

ATP consumption rate: type *1* > type 2 > type 3

From this order, one would expect that resistance to fatigue is highest for type 3, lowest for type 1, and intermediate for type 2. In our first study, Smith and I saw clear indications that this is indeed the case.

Factors in muscle fatigue

Fatigue studies on single *Xenopus* fibres are now being performed by both Elzinga and myself. In my laboratory, we are using mainly the very short fibres from lumbrical muscles (to facilitate microelectrode recordings). Both continuous and repetitive stimulation patterns have been used, mainly the latter. Figure 12.3A, illustrating an SDH-stained cross-section of a bundle of lumbrical fibres, shows that in these muscles as well, fibres of different kinds are found. Figure 12.3B shows average fatigue curves, obtained with repeated tetanic stimulations. Three groups of fibres are clearly evident, which we tentatively equate with types 1–3 in the iliofibularis muscle.

Fatigue curves of the type shown in Fig. 12.3 agree quite well with

12.3. (A) Cross-section of a bundle of lumbrical fibres from *Xenopus,* stained for succinate dehydrogenase (SDH) activity. Section prepared in collaboration with W. J. van der Laarse. (B) Average fatigue curves for altogether 37 single lumbrical fibres. Fibres are typed on the basis of fatigue resistance. Values are expressed as means ± SD. Stimulation duration 500 ms; 70-Hz tetanus, initially at 0.3 Hz and stepwise increased to 0.8 Hz (at 10–12 min stimulation time), after which tetanus duration was increased (final value, 800 ms).

A

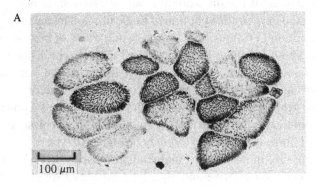

B

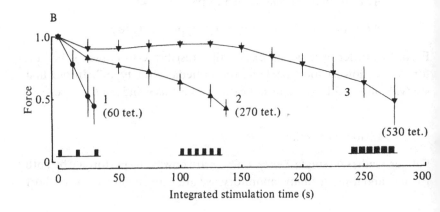

what can be predicted from the calculated rates of ATP formation and usage. However, they give little information about the process, or the factor, that is limiting when fatigue develops. Probably, the dominating theory is that high concentrations of metabolites, especially hydrogen ions and inorganic phosphate, interfere with cross-bridge activity and thus are the main cause of fatigue. This theory is to a large extent based on results from skinned fibre experiments (e.g., Cooke et al., 1988; Godt & Nosek, 1989). Our own results from intact fibres do not support this idea (Lännergren & Westerblad, 1989). Our main finding in this connection is that fibres that have been fatigued to about 0.4 P_0 – either with about 50 tetani (type 1) or about 500 tetani (type 3) – give essentially undiminished tension responses to caffeine application, as illustrated in Fig. 12.4.

Fig. 12.4. Tension responses to the application of caffeine at the end of fatiguing stimulation of a type 1 fibre (A) and a type 3 fibre (B). In (A), the whole stimulation period is shown (every fifth tetanus); in (B), the gap represents 580 intervening responses. [B reproduced by permission from H. Westerblad & J. Lännergren (1987). *Acta Physiologica Scandinavica, 130*, 357–8.]

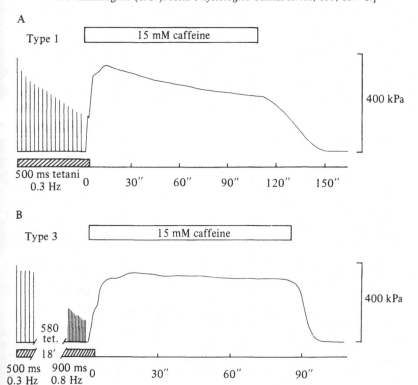

Thus, under our conditions at least, failing excitation–contraction (e–c) coupling seems to be the major cause of fatigue. More direct evidence for this view has recently been obtained by D. G. Allen, Lee, and Westerblad (1989), who recorded calcium transients in aequorin-injected lumbrical fibres during fatiguing stimulation. Concomitant with the fall in force, there was a decline in amplitude of the calcium transient which could largely explain the tension loss. Evidence for an additional factor, a decrease in calcium sensitivity at the contractile proteins, was obtained in a later study, where fura-2 was used as a Ca indicator (Lee, Westerblad, & Allen, 1991).

With evidence supporting failing e–c coupling as a major factor in muscle fatigue, an important question is, How is calcium release coupled to or influenced by metabolism? A clear answer to this question, it would seem, will have to await progress in studies of purified calcium-release channels. In the meantime, there are many other problems related to energy turnover in which *Xenopus* fibres, with their great range of metabolic and contractile properties, might be useful.

13

An electron microscopist's role in experiments on isolated muscle fibres

LUCY M. BROWN

Introduction

"Everything you see in the electron microscope is an artefact"! This truism does not mean that observations on denatured cells are a waste of time. The information has to be corroborated by other methods, preferably with observations on fresh tissues that are still able to perform their chemical and mechanical functions. When Andrew Huxley came to University College London 1960, he already had experience using electron miscroscopy (A. F. Huxley, 1959). I took charge of the Siemens EM1 previously used by Hugh Huxley who was moving to Cambridge, and I used it for most of the work described in this review.

Andrew Huxley's insight into problems of light and electron microscopy was an enormous help to me. In his B.Sc. class, he demonstrated the role of diffraction in the formation of the light microscope image, using striated muscle as a grating and a moveable stop in the back lens of the objective. When I was using laser diffraction to measure the periodicities in electron micrographs, he insisted that I understand the errors that could occur (Brown, 1975a). His pursuit of accuracy was insatiable. We once spent six weeks trying to take a light micrograph of a cross-hatched grating replica with oblique illumination. The spacing of 0.45 μm was within the limits of resolution but we were frustrated by the presence of lens aberrations (coma). He wasn't averse to buying expensive equipment when it was necessary, but found inexpensive solutions to many technical problems. Light micrographs were taken with the camera suspended over the microscope. When both were focussed at infinity, all problems with vibration and stray light were avoided.

Andrew Huxley's main contribution to electron microscope (EM) technology was the design of the Huxley microtomes, Marks 1 and 2. These cut excellent sections, were inexpensive and required very little maintenance. He introduced a four-hole condenser aperture into the Siemens EM1 to facilitate focussing at low magnifications and devised the accurate calibration described below.

Working with Andrew Huxley or his associates, I was involved in experiments on skeletal muscle fibres isolated from the frog, until my retirement in 1985. The experiments fell mainly into two groups, both of which entailed comparison with fresh fibres and accurate measurements of electron micrographs to determine the longitudinal alignment of the filaments. The first group of experiments was concerned with the appearance of wavy (unstimulated) myofibrils, seen first in fibres stimulated asymmetrically (A. F. Huxley & Gordon, 1962), and thereafter in passive fibres made wavy by longitudinal compression, which Hugo Gonzalez-Serratos had used at UCL for investigating the inward spread of activation (Gonzalez-Serratos, 1971). Collaboration with Gonzalez-Serratos was carried on by post from Mexico. We had some difficulty in persuading Huxley that muscle fibres in the passive state were interesting, but once he was convinced, he spent a fortnight in Mexico on some crucial experiments on fresh fibres (Brown, Gonzalez-Serratos, & Huxley, 1984a,b).

The second series of experiments was performed with Lydia Hill between 1975 and 1982 at UCL and Charing Cross Hospital Medical School (Brown & Hill, 1982, 1991). We followed by Lydia's observations made on fibres stretched during a tetanus when she was Andrew Huxley's Ph.D. student (Hill, 1977). The controls for these studies also shed light on the problem of the anomalous tension on the descending limb of the length–tension curve known as "creep".

Over the years, Huxley's associates dissected branched muscle fibres, and I examined some of them in the electron microscope (Brown et al., 1982).

Technical considerations

Fixation. It has not been possible to preserve the alignment of the filaments in both passive and actively-contracting single fibres with a single fixative. Glutaraldehyde in phosphate buffer was suitable for fixing passive wavy fibres (Brown et al., 1984a). If a fibre was damaged, the fixative stimulated it and the waves disappeared. Fixing active fibres was more difficult. Andrew Huxley and I monitored the fixation of fibres with a contracture at one side (A. F. Huxley & Gordon, 1962). We were successful in fixing one fibre with osmium, but we were unable to repeat the experiment because the whole fibre usually contracted so that the local contracture disappeared. Glutaraldehyde or formaldehyde caused the fibre to relax.

Tetanised muscle has been fixed with osmium by Page and Huxley (1963) and an unstable mixture of osmium and glutaraldehyde by Bergman (1983). Rapid freezing (Somlyo et al., 1981) avoids the possible artefacts introduced by fixation, but substitution with acetone does not avoid shrinkage, and only small areas of the muscle are preserved.

Lydia Hill and I confirmed that glutaraldehyde was unsuitable for single tetanised fibres because the tension was not maintained and the fibres remained excitable for several seconds. Andrew Huxley suggested Carnoy's method, which had been used for muscle fixation by Frank (1928). We (Brown & Hill, 1982) found that mercuric chloride in ethanol and chloroform fixed the protein elements, but not the membranes, and fulfilled our criteria for rapid fixation: (1) a large proportion of the tetanic tension remained after 15 min of fixation; (2) the band patterns were similar across the diameter of a fibre; and (3) the shrinkage artefact was comparable to standard methods. The fixative stimulated passive fibres so that the latter could not be used as controls. The method was subsequently used by Street (1983) as an extractive fixative to demonstrate transverse elements of the cytoskeleton in skeletal muscle.

The shrinkage artefact. Accurate measurement of sarcomere and filament length is prejudiced by the shrinkage that occurs during preparation for electron microscopy (A. F. Huxley & Peachey, 1961; Page & Huxley, 1963). A correction was obtained from the filament periodicity of fresh muscle measured in x-ray diffraction patterns (H. E. Huxley & Brown, 1967). The 14.3-nm period provided an index of thick-filament shrinkage. The best estimate for the troponin period is 38.9 nm – the mean of the doublet at 38.3 and 39.5 nm (Haselgrove, 1975). Measurement of the troponin periodicity in electron micrographs by laser diffraction (O'Brien, Bennett, & Hanson, 1971) was accurate to about 1% (Brown et al., 1984b).

In sarcomeres with no H-zone, thin filament length was estimated from the number of periods. The best estimate for the thin-filament–Z-line complex of frog skeletal muscle is 1.975 μm. Direct measurements corrected for shrinkage give a slightly lower value, probably because the last troponin gives a sharp edge to the H-zone. There is no evidence that the number of periods varies between different fibres in the frog, but there may be differences in Z-line width.

Fig. 13.1 shows that, in rapidly fixed tetanised fibres, the situation was complicated by shrinkage in the overlap zone, and slight stretching in the I-band to give a troponin period of up to 40.2 nm (Brown & Hill, 1982). Page and Huxley (1963) obtained similar values in tetanised mus-

Fig. 13.1. Cross-bridge and troponin periodicities in a tetanised fibre fixed with mercuric chloride in ethanol and chloroform. (A) Electron micrograph of a half-sarcomere. (B) Recombined images of the H-zone, overlap zone, and I-band. (C) Optical diffraction patterns; scale bar: 1 μm; periodicity: in nanometers (nm).

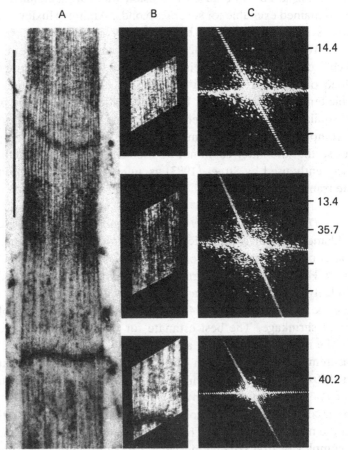

cle fixed with osmium. This suggests that the filaments were fixed in their correct position and did not pull out of the overlap zone. Shrinkage occurs during dehydration with ethanol or acetone. A water-soluble embedding medium might avoid these problems altogether.

Calibration of electron micrographs. There are three major sources of error: (1) variations in the distance between the specimen and the objective lens, (2) hysteresis in the intermediate and projector lenses, and (3) elliptical distortion. Pin-cushion and spiral distortion near the centre of the field are less serious. Andrew Huxley devised a method for calibrating the Siemens EM1 to an accuracy of 1% by using calibrated field-limiting apertures in the intermediate screen and the projector lens (Brown 1975b). The objective magnification was found by comparing spots of dirt on paired micrographs taken in the light and electron microscopes. A similar degree of accuracy can be obtained from a microscope with a eucentric goniometer – for example, the JEOL 100C (Brown, 1978) – by bringing the specimen to the eucentric point and inserting a calibrated selected area (SA) aperture into the objective image plane. I am greatly indebted to Andrew Huxley and to the late R. Willis for their helpful suggestions during this project.

Wavy myofibrils

Wavy myofibrils in fibres with a contracture on one side were detected in the interference microscope (A. F. Huxley & Gordon, 1962). Andrew Huxley and I confirmed that whereas sarcomeres with normal A- and I-bands were present on one side (Fig. 13.2A), there were contraction bands on the other (Fig. 13.2C). C_m bands are formed where the thin filaments overlap across the M-line, and C_z bands are formed where the thick filaments reach the Z-line (H. E. Huxley, 1965). The appearance of wavy myofibrils with A- and I-bands in the centre of the fibre suggested that they were shortened passively and that the true sarcomere length measured parallel to the myofibrils (Fig. 13.2B) might be much longer than the striation spacing, S.

Examination of wavy myofibrils in passive fibres prepared by Hugo Gonzalez-Serratos (1971) by longitudinal compression in gelatine confirmed that the sarcomeres resembled those in fibres at slack length (Fig. 13.2D) and led to the hypothesis that the waves appeared when the thin filaments of opposite polarity met in the M-line (Brown, Gonzalez-Serratos, & Huxley, 1970). Subsequent work on fresh and fixed fibres

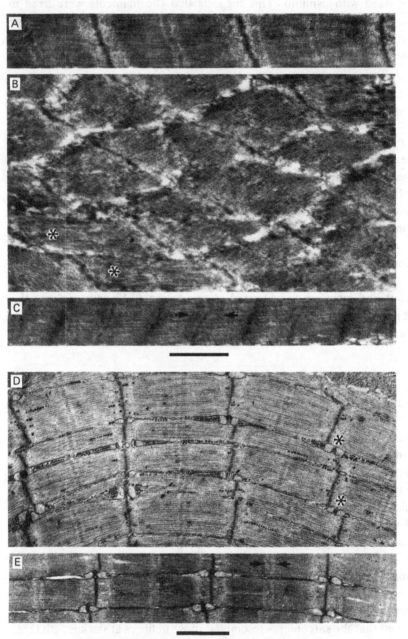

Fig. 13.2. Wavy myofibrils. (A–C) Fibre with a contracture on one side, fixed with osmium and stained with phosphotungstic acid. (A) Unstimulated myofibrils. (B) Wavy myofibrils; asterisks indicate sarcomeres in the plane of

showed that this was not so (Brown et al., 1984a,b). There were small differences between (1) S at slack length, (2) the critical S at which waves first appeared, and (3) the final S after compression. The position of the thin filaments at the M-line varied between fibres, with a small H-zone visible in several (Fig. 13.2E). The thin filaments overlapped slightly at the inside of bends in many cases. There is, therefore, no obvious structural barrier to shortening.

The force that resists longitudinal compression and leads to buckling of the myofibrils can be identified as the same force that restores a fibre to its slack length after active shortening (Ramsey & Street, 1940). The origin of the force, and calculations of the bending stiffness of the myofibrils and the filaments, are discussed in our second paper (Brown et al., 1984b).

Variations in filament overlap during a tetanus

Tension creep. The cross-bridge model of force generation (A. F. Huxley, 1957a, 1980; Gordon, Huxley, & Julian, 1966b) was rejected by Pollack and colleagues (ter Keurs, Iwazumi, & Pollack, 1978; Pollack, 1983) on the grounds that, in isolated muscle fibres at S longer than 2.2 μm, the tension creeps up during a prolonged tetanus so that the length–tension relation is not linear. A. F. Huxley (1988) has reviewed the evidence that the discrepancy is due to variations in filament overlap, and hence in the number of available cross-bridges, at any given S. Numerous studies of isolated fibres using light microscopy and laser diffraction have demonstrated nonuniformity of S, both near the tendons (A. F. Huxley & Peachey, 1961; Julian, Sollins, & Moss, 1978; Altringham & Pollack, 1984; Altringham & Bottinelli, 1985) and in different regions along the fibres (Gordon et al., 1966b; Julian & Morgan, 1979a; Edman & Reggiani, 1984).

Caption to Fig. 13.2. *(cont.)* the section. (C) Contracture showing C_m bands (arrows) and C_z bands. (D, E) Fibres shortened passively in gelatine; fixed with glutaraldehyde; and stained with osmium, uranyl acetate, and lead. (D) Fibre with no H-zones; the light lines on each side of the M-line mark the "pseudo-H-zone" where there are no cross-bridges. The Z-lines at the crest of the wave run perpendicular to the long axis whereas the Z-lines at either side are oblique to it, forming wedge-shaped sarcomeres with shorter sarcomere lengths at the inside of bends. Asterisks indicate discontinuities in the Z-lines of adjacent myofibrils. (E) Fibres with narrow H-zones. The ends of the thin filaments appear as a sharp boundary (heavy arrows), whereas the M-lines are irregular. Scale bars: 1 μm.

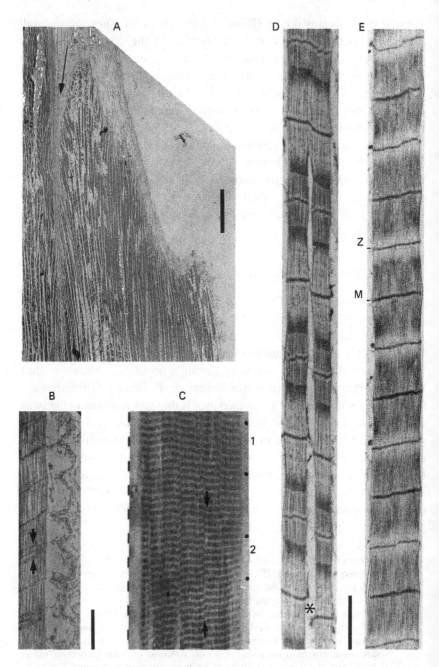

Fig. 13.3. Variations in filament overlap in tetanised fibres. The fibres in Figs. 13.3–13.5 were fixed with mercuric chloride in ethanol and chloroform. (A, B) A tendon insertion. (A) Low-magnification EM shows a cleft contain-

Huxley has pointed out that nonuniformities could occur that would not be detected by laser diffraction (A. F. Huxley, 1980a,b). Lydia Hill and I (Brown & Hill, 1982, 1991) have used polarised light with compensation and electron microscopy, to study isolated muscle fibres fixed rapidly during a tetanus (Fig. 13.3). We have confirmed that there was shortening near the tendons: C_m contraction bands and loops in the sarcolemma were seen close to, but not at the extreme tip of a tapering tendon insertion (Fig. 13.3A & B). S varied randomly along the fibres by up to 10%, and in complete vernier formations and smaller areas (Fig. 13.3C).

Electron microscopy revealed (1) variations in the width of the overlap zone within individual sarcomeres and (2) differences among sarcomeres across the diameter of a fibre. In a pilot experiment (Fig. 13.3D & E), we found a significant difference in overlap within the sarcomeres, between a fibre tetanised at $S = 2.8$ μm exhibiting creep of tension and one without creep at $S < 2.0$ μm. Measurements between the M- and Z-lines were made at intervals across half-sarcomeres (Brown & Farquharson, 1984). The variation in filament overlap is given by the SD of the mean M–Z distance about the mean for each half-sarcomere.

Single tetanised fibres have not previously been studied in the electron microscope. Many studies of muscle contracting under various conditions have demonstrated variability of sarcomere length and filament overlap at S longer than 2.2 μm (Page & Huxley, 1963; H. E. Huxley, 1965; Eisenberg & Eisenberg, 1982; Bergman, 1983; Horowits & Podolsky, 1987; Tsukita & Yano, 1988). Somlyo et al. (1981) describe only slightly disordered filaments in small bundles of tetanised frog fibres rapidly frozen at $S = 2.8$ μm. Their records show no creep of tension during a 1.2-s tetanus. This result is not inconsistent with our hypothesis that creep is associated with an increase in nonuniformity. Examination of single fibres rapidly fixed or frozen at different times during a tetanus and over a range of S would resolve this problem.

Caption to Fig. 13.3. *(cont.)* ing collagen (arrow). The tip of the longer branch with A- and I-bands (on the left) is not shown. Sarcomeres with C_m bands extend about 150 μm from the tip of the shorter branch. Scale bar: 10 μm. (B) C_m bands (arrows) and loops in the sarcolemma attached at the level of the M- and Z-lines. Scale bar: 1 μm. (C) One-micrometer section of a fibre viewed in polarised light with compensation; (1) $S = 2.94$ μm; (2) $S = 2.63$ μm; arrows indicate vernier formation with 16 striations on one side and 17 on the other. Graticule divisions: 10 μm. (D) Fibre tetanised at $S = 2.6$–2.9 μm. The branching myofibril gives rise to a vernier formation (asterisk). The M-lines are not parallel to the Z-lines, and the edges of the A–I junction are indistinct. (E) Fibre tetanised at $S = 1.7$–2.3 μm. The M- and Z-lines are more regular than they are in (D). Scale bar: 1 μm.

Stretch. When a frog fibre is stretched during a tetanus at S longer than 2.2 μm, there is an initial sharp rise and fall in tension, followed by a plateau (Fig. 13.4). This long-lasting residual tension ("component 3"; Edman, Elzinga, & Noble, 1984) is much higher than it would be in a fibre tetanised at the final length. Abbott and Aubert (1952) and Julian and Morgan (1979b) believed that nonuniformity of S was responsible. However, Hill (1977), using the interference microscope, found that the A-band width was constant and that S was more stable after stretch, and Edman, Elzinga, and Noble (1978, 1982, 1984) could not detect any nonuniformity by laser diffraction.

Lydia Hill and I (Brown & Hill, 1991) have examined three such fibres in the electron microscope after rapid fixation (Fig. 13.5). All showed a plateau of residual tension that was as high as, or higher than, the initial tension. In two that were stretched by 10% of their length at P_{max}, we found greater variations in overlap than in tetanised fibres, without S appeared perfectly regular in polarised light. At relatively short S, there were sarcomeres with very irregular M-lines and a few with C_m bands (Fig. 13.5A & B). At $S = 2.8–3.1$ μm, a few sarcomeres were elongated and disordered (Fig. 13.5C & D). In the fibre stretched by 17%, both polarised light and electron microscopy revealed gross irregularities (Fig. 13.5E).

The pattern of scattered areas of disordered sarcomeres in Fig. 13.5C is similar to that seen in human muscle damaged after prolonged eccentric contractions (Newham et al., 1983; Fridén, 1984). Alternatively, the disorder may be temporary, and the filaments slide back and regain their normal alignment after relaxation.

Discussion

Rapid fixation of single fibres with mercuric chloride in ethanol and chlorform has demonstrated that nonuniformity of filament ovarlap is present during a tetanus and may be aggravated by stretch. At S longer than 2.2 μm, the tension is higher than predicted from the length–tension curve. Our measurements of filament overlap in tetanised fibres, and the observations on fibres stretched during a tetanus, are pilot experiments only. An extensive study would be necessary to demonstrate that the number of available cross-bridges had increased sufficiently to account for all the anomalous tension. This might be feasible in tetanised fibres but would be well nigh impossible in stretched ones. In the fibre stretched by 17%, there would have to be sufficient sarcomeres that had shortened below their initial length, and areas of increased overlap in the grossly elongated sarcomeres.

Hill (1977) and Edman et al. (1978) found that tetanised fibres showed no residual tension enhancement after stretch if S did not exceed 2.2 μm. Nonuniformity would not be expected to generate extra tension above P_{max}, but recently a small increase in the residual tension has been observed (Colomo, Lombardi & Piazzesi, 1988; Colomo et al., 1989a). Colomo et al., (1988) and Sugi and Tsuchiya (1988) showed that stiffness decreased relative to the residual tension enhancement in fibres after slow stretches. If stiffness is a measure of attached, as distinct from available, cross-bridges, it suggests that nonuniformity cannot completely account for the anomalous tension and that other factors may be involved.

Sugi and Tsuchiya suggested that stretch might cause disorder in the filament lattice. The intensity of the [1,1] reflection in the x-ray diffraction pattern decreased during stretch and then partially recovered during the remainder of the tetanus (Amemiya et al., 1988). Edman et al. (1978, 1982, 1984) suggested that a parallel elastic element might contribute to the tension enhancement, and recently, Tsuchiya and Edman (1990) have obtained further evidence for it.

There may be both longitudinal and transverse forces that develop in the filamentous cytoskeleton during a prolonged tetanus, or stretch during a tetanus. Connectin (titin) filaments run from the thick filaments to the Z-line (Maruyama, 1986), and there is evidence that these may keep the thick filaments aligned during passive stretch and recentre them after tetanic contraction (Horowits et al., 1986; Horowits & Podolsky, 1987, 1988). The network of filaments containing desmin (Nelson & Lazarides, 1984), illustrated in Fig. 13.3B, may maintain the transverse

Fig. 13.4. Tension record of the tetanised fibre illustrated in Fig. 13.5C and D. Small arrow; fibre stretched by 10% of its length at P_{max}. Large arrow: application of the fixative; about half the tension remained after 15 min. Time trace: seconds.

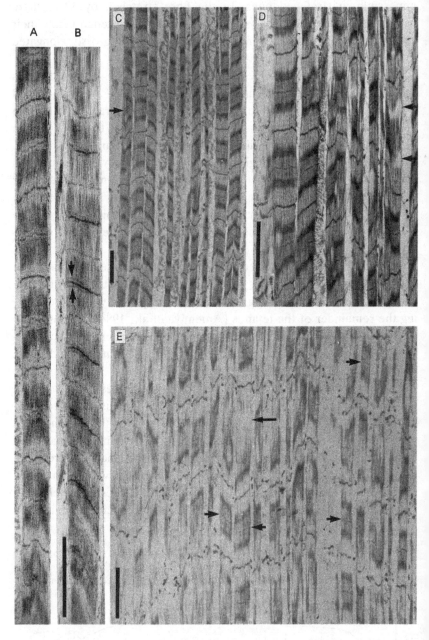

Fig. 13.5. Three fibres stretched during a tetanus. (A, B) Fibre stretched by 10% to $S = 2.2$–2.5 μm with very irregular M-lines and a few C_m bands (arrows). (C, D) Fibre stretched by 10% to $S = 2.8$–3.1 μm. (C) Most

alignment of the M- and Z-lines so that there is only moderate displacement in vernier formations (Fig. 13.3C) and between wavy myofibrils (Fig. 13.2D). In stretched fibres the displacement of the M- and Z-lines is more pronounced. Street (1983) has shown that both active and passive tension can be transmitted laterally, presumably through the transverse cytoskeleton and the extensive branching of the myofibrils (Fig. 13.3D).

Morgan (1990) has modelled a muscle fibre stretched during active contraction and has found a pattern of nonuniformity very similar to ours. Individual sarcomeres, randomly distributed, undergo rapid uncontrolled lengthening so that the filaments no longer overlap. In our case, the excessive elongation may be confined to groups of filaments within a half sarcomere, but the mechanical effects should be similar. The tension in the "popped" sarcomeres is borne by passive components. The majority of the sarcomeres do not lengthen, which explains why the residual tension is at least as high as the initial tension. The behaviour of the model during active stretch depends on the presence of slight nonuniformities before stimulation.

Morgan's model provides an explanation for our preliminary observations on active lengthening. It now seems clear that the anomalous tension recorded after both creep and stretch during a tetanus is due mainly to nonuniformity of filament overlap. More experiments are needed to ascertain whether unlengthened sarcomeres can account for all the extra tension recorded in actively lengthened fibres. Our suggestion that the degree of nonuniformity in tetanised fibres is related to sarcomere length also needs to be confirmed.

Caption to Fig. 13.5. *(cont.)* sarcomeres have relatively regular overlap zones. Arrow indicates sarcomere elongated on one side. (D) A small area with several disordered sarcomeres. Arrows indicate gaps between the thick and thin filaments. (E) Fibre stretched by 17% to $S = 2.8$–4.5 μm with numerous disordered sarcomeres. Long arrow: half sarcomere with no overlap. Short arrows: sarcomeres with relatively regular overlap zones. Scale bars: 2 μm.

14

Structural changes accompanying mechanical events in muscle contraction

ROBERT M. SIMMONS

Introduction

My first degree was in physics, but I became interested in the structure of biological macromolecules in my final year, and went on to do a Ph.D. in protein crystallography. I found it very frustrating having to wait for crystals to grow and decided to change to a research area where I could go into the laboratory in the morning and do a new experiment. The easiest option was to change to fibre diffraction, and muscle seemed a good thing to work on as its structure was less well worked out than many of the other fibrous proteins. However, somewhere along the way to securing a postdoctoral position, I was given an introduction to Andrew Huxley, then head of the Department of Physiology at University College London (UCL), who persuaded me to think seriously about a career in physiology. So I did an M.Sc. in Physiology at UCL and ended up doing a project in Prof's laboratory. My project was to investigate a recent report that the internal resistivity of muscle fibres changed during contraction (it didn't).

After that I joined Prof in his research on the mechanical properties of muscle fibres, taking over where Fred Julian had left off (see Chapter 2). It took several years to make the length step experiments on muscle fibres

I wish to thank all the colleagues who have collaborated with us to make these experiments possible, especially Y. Maeda (EMBL, Hamburg), J. Bordas and E. Towns-Andrews (SERC, Daresbury), C. Bond (MRC, Cambridge), and K. Eason (MRC, London).

that led to the cross-bridge theory paper in 1971 (A. F. Huxley & Simmons, 1971b), and the really quantitative experiments were not done until Lincoln Ford arrived. After that, Prof was busy with the analysis of data, and I decided to have a look at the mechanics of skinned fibres – in particular, to see what happened when the concentrations of ATP and its products were changed. Fortunately, Yale Goldman came to work with Prof about this time and joined in on this project, and together with Michael Ferenczi got the methodology worked out (Ferenczi, Goldman, & Simmons, 1984; Goldman & Simmons, 1984). Over the ensuing years, I became steadily more convinced that we needed to combine mechanics with structural techniques if we were ever to understand what the cross-bridges were doing, and in 1979, I joined forces with Hugh Huxley and his group working at the Hamburg synchrotron. Thus I finally did some fibre diffraction work, albeit on a part-time basis because of the scarcity of beam time. This led to a collaboration with Bernhard Brenner to work out the best conditions for doing single-fibre research using synchrotron radiation. Recently, some of the x-ray results reminded me of one of the earliest experiments I did with Prof, on the "pull-out" effect, and Kevin Burton and I have taken that a bit further in some experiments reported here.

X-ray diffraction pattern of frog skeletal muscle (collaborators: H. E. Huxley and A. R. Faruqi; B. Brenner)

The x-ray diffraction pattern is obtained simply by placing a muscle in a focussed beam of x-rays. With synchrotron radiation the beam is typically a few millimeters across at the specimen and a millimeter or less at the focal point, which is set at 1–6 m beyond the specimen, depending on the resolution required. For fast time-resolved experiments the pattern is usually recorded with the use of a position-sensitive detector.

Most of the x-ray diffraction pattern arises from the regular arrangement of myosin molecules along the thick filament and of actin molecules along the thin filament (Fig. 14.1; reviewed by Squire, 1981; Haselgrove, 1983). The equatorial pattern arises from the projection of the structure down the fibre axis, and in the experiments described here it is mostly dependent on the hexagonal array of the filaments in the overlap zone. It is dominated by two strong inner reflections, [1,0] and [1,1]. For reasons of geometry, the intensity of both of the reflections depends on the ratio of the masses associated with the thick and the thin filament, respectively, but the ratio is larger for the [1,0] reflection than for the [1,1] reflection. In a relaxed muscle, the intensity of the [1,0] reflection (I_{10}) is larger than the intensity of the [1,1] reflection (I_{11}), so that $I_{10} > I_{11}$, as most of the cross-bridges are close to the thick filament backbone; in contrast, in rigor, $I_{10} < I_{11}$, as most of the cross-bridges are attached to actin sites (Haselgrove & Huxley, 1973). At the plateau of an isometric tetanus, $I_{10} < I_{11}$, but the extent of the change from the relaxed pattern is somewhat less than in rigor, probably indicating that fewer than the maximum possible number of cross-bridges are attached at any one time. The change in the ratio from the relaxed to the contracting state is shown in Fig. 14.2, which is from an early experiment of Hugh Huxley's, with a position-sensitive detector placed along the equator. I_{10} and I_{11} change slightly ahead of tension during the rise of tension in a tetanus (H. E. Huxley, 1979; H. E. Huxley et al., 1980; Fig. 14.5). So does stiffness (Cecchi, Griffiths, & Taylor, 1982, 1986; Ford, Huxley, & Simmons, 1986), and both of these effects probably result from cross-bridges initially attaching to actin in a "preforce" state and then making a transition to a force-producing state. The meridional pattern is similarly dependent on the regular repeat of the structure along the axis and is dominated by the myosin cross-bridge repeat of 14.3 nm (though there are many other weaker reflections on the meridian that are not dealt with here). During an active contraction, the intensity of the 14.3-nm reflection ($I_{14.3}$) increases with about the same time course as the tension

Fig. 14.1. X-ray diffraction pattern of resting frog sartorius muscle (A) and its interpretation in terms of layer lines, etc. (B). The equatorial reflections cannot be seen in (A) because the film has been overexposed (see Fig. 14.2). Units: Å. [Reproduced by permission from J. Squire (1981). *The Structural Basis of Muscular Contraction.* New York: Plenum press. The x-ray diffraction pattern in A was taken by Dr. J. Haselgrove and is reproduced with his permission.]

A

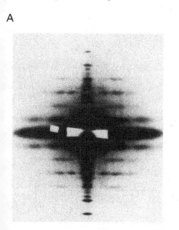

B

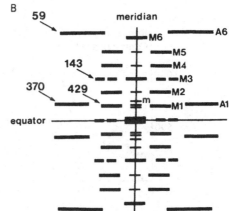

(see, e.g., the start of the record in Fig. 14.6A, though the time course is often more complicated), so presumably the ordering of the cross-bridges improves as they bind to actin. There is also an increase in the spacing of the (nominal) 14.3-nm periodicity ($d_{14.3}$) by about 1% (Fig. 14.6B) and this increase also occurs in rigor. The reason for this change is not clear. It may occur as a result of a change in length of the thick filament or other structure (reviewed by H. E. Huxley, 1979; Haselgrove, 1983). Alternatively, attached cross-bridges may have an intrinsically different periodicity from detached cross-bridges, perhaps because they now follow the nearest thin filament periodicity which at this level may be closer to 14.5 nm (cf. Squire, Luther, & Morris, 1990). Because of problems of detector instability in the past, it is only recently that we have been able to record the time course of $d_{14.3}$ accurately, although Huxley and Murray (H. E. Huxley, 1979) showed that it started to change ahead of tension at the start of a tetanus.

The remainder of the relaxed pattern is dominated by the layer lines from the helical array of the cross-briges, based on the total repeat of 42.9 nm. In rigor these layer lines are replaced by another set based on the thin filament repeat of 36.0 nm, as the cross-bridges enhance the thin filament structure. In an active contraction the layer line pattern becomes very weak, presumably because the cross-bridges are disordered in a radial direction. (There is a marked contrast between the precise mechanical behaviour of an active muscle and its apparently formless structure, and

Fig. 14.2. Equatorial diffraction pattern of frog sartorius muscle in the resting state and during the plateau of an isometric tetanus. Unpublished data of H. E. Huxley, using a position-sensitive detector with a rotating anode x-ray generator. (Reproduced by permission of H. E. Huxley.)

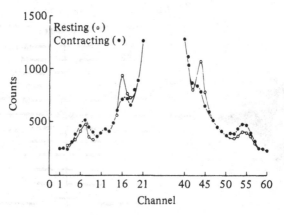

as a consequence it has been impossible to determine which part of the pattern corresponds to the mechanical behaviour and which part to uninformative disorder. However, in recent work at the Daresbury synchrotron by Joan Bordas and his colleagues and by ourselves, using a two-dimensional detector with good spatial resolution and efficiency, there are encouraging signs that it may be possible to "unscramble" the pattern.)

The results presented thus far represent substantially what was known in 1979 about the time course of x-ray diffraction changes at the beginning and during a tetanus, at the time when we decided to try to record the pattern during rapid transient experiments. We wanted to see in particular if there were changes corresponding to the very rapid (millisecond time scale) transients which A. F. Huxley and Simmons (1971b) had ascribed to the force-generating step of the cross-bridges. In comparatively crude experiments (using whole muscles, which are needed to give adequate time resolution, with no sarcomere length control), we managed to achieve a useful time resolution of 1 ms in the most intense parts of the pattern (H. E. Huxley et al., 1981, 1983). However, the only major changes we found were on the meridian – $I_{14.3}$ showing a dramatic decrease in near zero for sufficiently large stretches or releases (Fig. 14.3).

Fig. 14.3. Intensity of the 14.3-nm meridional reflection after a large release during a tetanus in frog sartorius muscle. □, $I_{14.3}$; △, tension. [Reproduced by permission from H. E. Huxley et al. (1983). *Journal of Molecular Biology*, *169*, 469–506.]

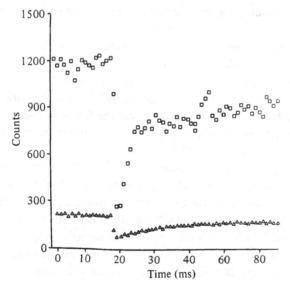

But it was difficult to say whether the changes took place synchronously with the length change or with a delay corresponding to the putative force-generating step. In the large releases the rapid transients were not visible on the time scale of our measurements. For smaller releases the changes in $I_{14.3}$ were also smaller and the signal/noise ratio was correspondingly worse. For stretches the transients are much slower, but we were unsure for steps large enough to see the changes in the x-ray pattern whether the filament sliding was within the normal range of cross-bridge action. The problems were compounded by effects due to changes of length in the tendons and the deterioration in the pattern that occurs at low temperature (but the transients are faster at higher temperatures!). Other changes on the meridian were smaller; for example, there was a transient increase (towards the relaxed value) in $I_{21.5}$ after a release. Changes in the intensity of the equatorial reflections were small and variable.

Interpreting these results requires some caution, but they seem to suggest that a rapid length change causes a nearly immediate disorientation of the attached cross-bridges from the regular array at 14.3 nm, perhaps indicating that all or part of the elastic element lies in the crossbridge head or at the actin site (and not in S2 as usually depicted; some alternative models are shown in A. F. Huxley, 1974). The disorientation implies that not all the heads are free to move in the same way, perhaps because some of the heads reach the end of their working stroke. It also implies, incidentally, that the cross-bridges must have been attached in the first place; it would be hard to explain the decrease in the $I_{14.3}$ on the basis of theories of contraction not involving cross-bridges. A partial recovery of the $I_{14.3}$ occurred at a rate intermediate between that of the rapid transients and the slower recovery back to the isometric level (Fig. 14.3), and this, coupled with the transient and partial return of the resting 21.5-nm reflection, indicates that many of the cross-bridges detach and reattach to actin sites further along the filament unexpectedly quickly. This explains in part why there is not much change in the intensity of the equatorial reflections.

These experiments were something of a landmark in time-resolved synchrotron radiation studies, but they perhaps raised more questions than they answered. Over the next few years we struggled to obtain better data, particularly for the equatorial reflections and also for the actin layer lines at 5.9 nm and 5.1 nm, but with the available sources and detectors we were clearly not making much progress. We eventually abandoned the attempt to obtain very fast time resolution data and turned to other experiments that required a lower time resolution. When the intensity of beam lines and the performance of detectors has been sufficiently improved, it will be necessary to return to the fast transient experiments, using a preparation in which sarcomere length control is possible. It was not until two or three years ago that we returned to the problem of cross-bridge dynamics. Now working at the SERC Synchrotron Radiation Source at Daresbury, we decided to examine what happens at the onset of shortening during the plateau of an isometric tetanus, arguing that one reason we had seen only small changes in the equatorial reflections with steps was that cross-bridges reattached too quickly. We would be better placed to look at the full cross-bridge cycle by forcing the cross-bridges to detach by more extensive sliding of the filaments. Previous experiments had already shown that in the steady state there was a partial return towards the resting pattern in muscles shortening at high velocity (H. E. Huxley, 1979), although the change at low and intermediate velocities was small (Podolsky et al., 1976). We anticipated also that there would be a change in $I_{14.3}$ on the basis of our step experiments, and Matsubara and Yagi (1985) had already detected such changes during shortening in laboratory experiments. We decided to do "simulated isotonic release" experiments, in which a step is first applied in a tetanus to reduce the tension rapidly, followed by a ramp to keep the tension at its new level (Fig. 14.4). We were surprised to find that, for the intensity of the equatorial

Fig. 14.4. Illustration of simulated isotonic release experiments. The muscle elasticity is released by a rapid length change taking 1–5 ms, followed by a slower shortening ramp to keep the tension approximately constant.

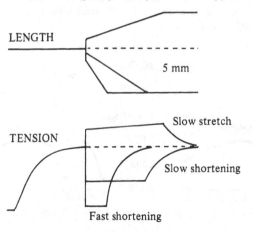

reflections and also for $d_{14.3}$, it took a long time (and distance of shortening) for the changes to reach a new steady-state level (H. E. Huxley et al., 1988; H. E. Huxley, Simmons, & Faruqi, 1989; Figs. 14.5 & 14.6). For shortening near the maximum velocity, it was not clear whether the steady state had been attained by the end of the movement, which was the maximum – about 12% or 100 nm of filament sliding – that our motor was capable of (we are now repeating the measurements using a motor with a longer throw). $I_{14.3}$ fell with no appreciable delay, as we had found before in the step experiments. At lower velocities the change in $I_{14.3}$ decreased roughly in proportion to velocity, but the changes in the equatorial reflections and $d_{14.3}$ were disproportionately small.

The extent of the return of I_{10} and I_{11} toward their resting values shows a much less steep dependence on shortening velocity than does the stiffness, which also gives information about the number of cross-bridges attached (Ford, Huxley, & Simmons, 1985). These two results, shown qualitatively but expressed commensurately, are shown in Fig. 14.7A. A similar discrepancy exists in rabbit skinned muscle fibres (Brenner, Maeda, & Simmons, 1990; Fig. 14.7B). There are numerous possible interpretations of this difference, but the simplest is that at slow velocities many of the cross-bridges are attached in a state that has little mechanical sign, perhaps corresponding to the weakly attached state of Eisenberg and Greene (1980). It should be noted that this state must be different from the "preforce" state implied to exist by both x-ray diffraction and stiffness measurements at the beginning of a tetanus, and which

Fig. 14.5. Intensity of the equatorial reflections during a contraction with imposed shortening near the maximum velocity: (A) I_{10}; (B) I_{11}. Ordinate: an arbitrary scale proportional to the total counts under peak without background subtraction. Frog sartorius muscle, 9°C; total distance of shortening, 5 mm; muscle length, about 40 mm.

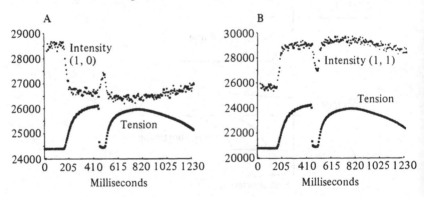

must contribute to stiffness but not to force. At higher velocities of shortening, cross-bridges do become detached as predicted by the theory of A. F. Huxley (1957a), because the rate of detachment becomes large as compared to the rate of attachment. It is also worth noting that there is nothing new about the implication that cross-bridges attach during shortening and are detached before releasing their products. Podolsky and Nolan (1971) introduced this concept to account for the observed ATPase rate of muscle during shortening. Further, even if the numbers of cross-bridges attached during shortening are assumed to be those implied from stiffness measurements, consideration of ATPase rates shows that at high velocities only a small proportion of the cross-bridges are committed to produce release, so that each cross-bridge would make perhaps ten transient attachments to each one that results in ATP hydrolysis (e.g., Bagshaw, 1982).

The dependence of $I_{14.3}$ on shortening velocity means that the mean deviation of axial cross-bridge locations increases with velocity. This was also predicted by the theory of A. F. Huxley (1957a). The change in $d_{14.3}$ with velocity is peculiar in two respects: One is the marked nonlinearity, which probably contributes to the apparent rapidity of the change at the start of a tetanus. The other is a definite delay after the tension falls. At present, we can only speculate about the reason for this, but it does rule out the possibility of a substantial contribution to the instantaneous compliance of the sarcomere from a length change in passive structures such as the filaments that could conceivably affect $d_{14.3}$. This supports the conclusion that most of the instantaneous compliance of the sarcomere

Fig. 14.6. Intensity and spacing of the 14.3-nm meridional reflection during a contraction with imposed shortening near the maximum velocity. (A) $I_{14.3}$; (B) $d_{14.3}$. Details as in Fig. 14.5.

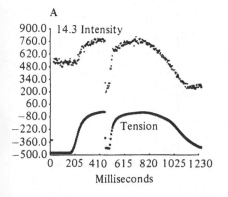

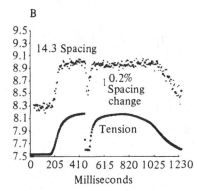

lies in the cross-bridges, based on the observation that fibre stiffness is proportional to filament overlap (Ford, Huxley, & Simmons, 1981).

One further point needs to be elucidated. This concerns the relation between the change of the equatorial reflection intensities and the change in $d_{14.3}$. It seems that the two may be connected in some way, but we do not yet know if the relation is a causal one (e.g., $d_{14.3}$ depends on the tension and therefore on the number of cross-bridges attached and consequently on the I_{11}/I_{10} ratio) or whether the two both depend on the same factor (e.g., attachment of cross-bridges to actin).

Future x-ray diffraction experiments

Before passing on to some further mechanical experiments on the rate at which structural changes occur during shortening, it seems a good opportunity to make some remarks on the difficulties of doing x-ray diffraction experiments. Unfortunately, muscle diffracts poorly (the concentration of protein is low), and, with current synchrotron radiation sources, which deliver about 10^{11} photons per second at the specimen, and with detectors that are 50% efficient or better, the total number of counts per second in various parts of the pattern are of the following orders of magnitude: equatorial (1,0) and (1,1), 10^5; meridional (14.3 nm), 10^4; layer line, 10^3. Using thicker muscles does not improve matters because

Fig. 14.7. Intensity of the equatorial reflections during steady shortening compared with stiffness. The difference in the value of I_{11}/I_{10} from its resting value has been divided by the corresponding value for isometric state (cf. Yu, Hartt, & Podolsky, 1979). (A) Frog sartorius muscle. Approximate changes in x-ray intensity based on Podolsky et al. (1976), H. E. Huxley (1979), and the present experiments. Stiffness values are based on Ford et al. (1985). (B) Rabbit psoas single muscle fibres, 5°C. (Based on unpublished data of Brenner & Simmons.)

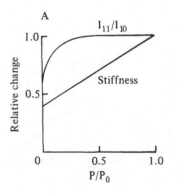
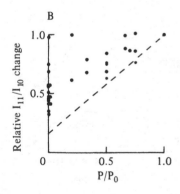

absorption becomes limiting beyond 1 mm. The noise in the signal is equal to the square root of the photon count, so that 10^4 counts are needed to obtain a noise level of 1%. For an equatorial reflection, the exposure needed is then 10^{-1} s, and if a time resolution of 10^{-3} is required, the contraction must be repeated 100 times. In fact, it is possible to get about 100 contractions each with an x-ray exposure of 1 s from a good muscle before it shows signs of beam damage or serious fatigue, so the data required can in principle be obtained from a single muscle. In general, the number of muscles needed is given by $1/(p^2ctn)$ where p is the fractional noise/signal ratio required, c is the number of counts per second, t is the time resolution, and n is the number of contractions per muscle. So the same quality of data would need 10 muscles for the 14.3-nm meridional reflection and 100 for a layer line. Of course, if the changes are large, the stipulation on the noise/signal ratio can be relaxed, and there are special cases in which it is possible to repeat an intervention several times within a single contraction. We are now convinced that the full two-dimensional pattern is needed to understand the structural changes, so that of the order of 10 muscles are needed to obtain data of even moderate quality (10% noise level for the weaker parts of the pattern) or moderate time resolution (10 ms). Allowing for setting up time and resting a muscle at least 3 min between each contraction, it takes about 6 h to collect the data from one muscle. In a week's work without mishap to the storage ring or beam line (which is rare), it is possible to run about 20 muscles. Thus it is usual for us now to collect data for two types of intervention (e.g., different velocities of shortening) in a week. The maximum amount of beam time allocated is about 4 weeks a year, of which about one-half to three-quarters is productive. Progress is therefore painfully slow. Higher-intensity sources or beam lines are needed to improve on this (provided radiation damage is not accelerated disproportionately).

Mechanical experiments on the onset of shortening
(collaborator: K. Burton)

Some recent results from simplified preparations have suggested that the cross-bridge working stroke is very much longer than it has usually been considered to be since the inception of the sliding-filament theory (e.g., Yanagida, Arata, & Oosawa, 1985). The working stroke became enshrined at about 10–15 nm, and there were historically a number of reasons for this. One was simply that the EM appearance of cross-bridges in rigor was the same, irrespective of the amount of sliding that had occurred; therefore, any change in structure producing the sliding must have been small. There were also more detailed arguments about the movement to be expected from the amount of ATP hydrolysed and the geometry of the filaments, which again gave small values for the working stroke (Hanson & Huxley, 1955; see Chapter 2, this volume). More detailed considerations of the mechanical and thermodynamic properties of the cross-bridge model were also consistent with a working stroke of about 10 nm (A. F. Huxley, 1957a). A value of 10–20 nm is also the maximum distance over which a tilting head could produce force (H. E. Huxley, 1969). Finally, a value of up to about 14 nm per half sarcomere is the distance over which very rapid filament sliding occurs when a muscle is rapidly shortened, after which it either shortens much more slowly or redevelops tension, depending on the conditions (e.g., A. F. Huxley & Simmons, 1971b).

However, none of these observations prove that cross-bridges cannot remain attached and produce tension, particularly at low levels, over a much greater distance. Also, a distinction has to be drawn between the elementary working stroke corresponding, for example, to the movement that an isolated myosin head would impart to an actin filament, and the distance over which the average cross-bridge would remain attached in a shortening muscle between successive cycles of ATP hydrolysis. As already mentioned in the previous section, it seems likely that cross-bridges make multiple interactions with actin sites between successive hydrolysis steps. The working stroke is thought to arise from a transition from a "preforce" weakly bound state to a strongly bound state (Eisenberg & Greene, 1980). A simple explanation for our observations on the rate of change of the x-ray diffraction pattern would be that after their working stroke, cross-bridges revert to a weakly bound state and remain attached for some time, perhaps until product release has occurred. However, the distinction between elementary working stroke and distance travelled per ATP hydrolysed can become almost entirely blurred, as it is only a short extrapolation to suppose that strongly bound cross-bridges can also slip from one actin site to the next (cf. Hill, 1974; Brenner, 1990).

It seemed puzzling, when our x-ray diffraction results were obtained, that there were no well-known comparable mechanical phenomena. There were some experiments of A. F. Huxley and Simmons (unpublished) in which stiffness was measured with an imposed oscillation during isotonic releases, and it seemed that the major part of the stiffness change occurred within a few milliseconds. The only effect that took some time to establish was the so-called pull-out effect, which occurs (for example) when one performs an isotonic release to a low tension and then reapplies the isometric force. Because the number of cross-bridges attached is reduced during shortening (or because they are less tightly bound), the muscle is extended extremely rapidly beyond its original length (Katz, 1939; A. F. Huxley & Simmons, 1973). The effect was studied in some detail by Sugi and Tsuchiya (1981), who showed that the initial velocity of lengthening increased as a function of time between the start of isotonic shortening and the application of a test load.

At Hugh Huxley's suggestion, we decided to investigate this phenomenon in a slightly different way, using the same step and ramp length changes used in the x-ray diffraction experiments followed by a test step to test the mechanical state of the cross-bridges, after which the fibre was kept isometric (Burton & Simmons, 1990; Fig. 14.8A). We used rabbit skinned fibres, in which this type of intervention had been introduced by Brenner (1983) to stabilise the sarcomere length distribution, using a restretch to bring the muscle fibre back to its starting length. The response to a restretch is a "kick" of tension followed by a minimum and then a slow redevelopment of tension (Fig. 14.8A). The filament sliding during the restretch can be of the order of 100 nm, many times the conventional cross-bridge range, and the response during the stretch is likely to be due to a viscouslike drag of cross-bridges being stretched

Fig. 14.8. Tension response to restretches applied during shortening. Rabbit psoas skinned muscle fibres, 5°C. (A) Restretch to initial length; shortening velocity, 0.38 lengths · s⁻¹; restretch from 1% to 5%; initial sarcomere length, 2.39 μm. (B) Restretch by a constant amount (3%); shortening velocity, 0.6 lengths · s⁻¹; initial sarcomere length, 2.34 μm. (A slower component of change in the tension response is present in (B), resulting from the change in the final length to which the restretch is made.)

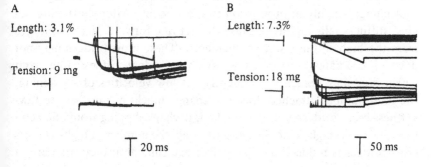

A

Length: 3.1%

Tension: 9 mg

B

Length: 7.3%

Tension: 18 mg

⊤ 20 ms   ⊤ 50 ms

until they detach (cf. Schoenberg, 1989). During lengthening, Vincenzo Lombardi and his colleagues have shown that cross-bridges that are forcibly detached are able to reattach rapidly (see Chapter 16, this volume; also Brenner, 1988b; Burton, 1989). The size of the tension kick and the minimum of tension after it are presumably measures of the number of cross-bridges able to attach during such fast lengthening. When the restretch is made soon after the onset of shortening, these measures are much higher than the corresponding steady-state levels, which are reached only after about 65 ms, corresponding to filament sliding of about 50–60 nm. This is, however, a confusing experiment to interpret since the size of the restretch varies, and in subsequent experiments we used a restretch of constant size. The results were similar if the restretch was large (>20 nm of filament sliding; Fig. 14.8B). On the other hand, we have subsequently shown that with small test stretches (<10 nm of filament sliding), which are within the range that give an elastic response (Ford, Huxley, & Simmons, 1977), the time course of the change of stiffness is much quicker and is complete within a distance of shortening corresponding to about 20 nm.

In order to explain both the mechanical and the x-ray diffraction results, at least in an ad hoc fashion and taking each at its face value, it must be supposed that there are two types of attached cross-bridge states – those with low stiffness and those with high stiffness. The low-stiffness states might correspond to the first and last attached positions of the cross-bridges, and the high-stiffness states to the intervening positions where there is a rapid equilibrium between a sequence of states producing successively more force (A. F. Huxley, 1980). The two types of states may be the same as the weak and strong binding states of Eisenberg and Greene (1980). We assume that in an isometric contraction the number of high-stiffness cross-bridges is relatively large, though possibly only a small proportion of the total number attached (H. E. Huxley & Kress, 1985), and this number decreases during steady-state shortening, roughly in proportion to the velocity. After shortening begins, stiffness falls rapidly because the net rate forward of cross-bridges from the high-stiffness state is increased. These cross-bridges may not detach immediately; instead they may enter the low-stiffness state, from which they may detach subsequently. At low velocities of shortening, the total number attached does not change much, as most of the low-stiffness cross-bridges remain attached, perhaps slipping from one actin to the next. At high velocities of shortening, the number of high-stiffness cross-bridges is reduced markedly. Net detachment also occurs but not rapidly, because low-stiffness cross-bridges (whether in the first or last position of attachment) have to complete their cycle before they finally detach. It has to be assumed that reattachment from the detached state is relatively slow.

This scheme is consistent with the idea that the mechanical response to a small restretch measures predominantly the number of cross-bridges in the high-stiffness state. It is then unlikely that cross-bridges remain attached to the thin filament during the working stroke (i.e., in the high-stiffness state) over the long distances proposed by Yanagida et al. (1985). The mechanical results for large restretches can probably be explained as the responses of all the cross-bridges (low- and high-stiffness) except those that have detached. The time course of the change towards the steady state is then relatively slow as compared to that with small steps, primarily because of the contribution of the low-stiffness cross-bridges.

15

Mechano-chemistry of negatively strained cross-bridges in skeletal muscle

YALE E. GOLDMAN

Introduction

This volume to honour Andrew Huxley presents an opportunity to review a line of investigation that was strongly influenced by his ideas and experiments. In fact, Andrew Huxley's leadership in studies of muscle contraction has been exceptional all the way from the recognition of the ionic nature of membrane events, through the inception of the sliding-filament theory and the proposition of cyclic cross-bridge interactions, to the consideration of mechanical constraints on enzymatic rate constants. I was a postdoctoral fellow in Physiology at University College, London, when Andrew Huxley was there. Although I was working with Bob Simmons during that fellowship, I was helped enormously by many conversations and interactions with Andrew Huxley. The experiments I summarise here followed from ideas he first expressed.

One of Huxley's earliest publications in the muscle field (A. F. Huxley, 1957a) was a brilliant paper reviewing studies on the contraction mechanism and presenting a novel hypothesis and quantitative model. In 1957, the idea of sliding filaments was new, and the nature of the interaction between thick and thin filaments was obscure. Based on the

This work was supported by the Muscular Dystrophy Associations of America and NIH grants HL15835 to the Pennsylvania Muscle Institute, and AR26846. I thank Dr. W. F. Harrington and my co-workers Marc Bell, Jody Dantzig, Alex Friedman, Hideo Higuchi, and Jonathan Tanner for helpful discussion of the manuscript.

linear fall of tetanic tension as striation spacing is increased (Ramsey & Street, 1940; A. F. Huxley & Niedergerke, 1954), and the biophysical properties of actomyosin threads, such as the increase of extensibility on addition of ATP (Weber, 1955), Huxley proposed that projections on the thick filaments undergo a cyclic interaction with the thin filament to transduce chemical to mechanical energy. In the mathematical description, specific assumptions were made for the rates of attachment and detachment of bridges between the thick and thin filaments and the dependence of those rates on the cross-bridge mechanical strain. Huxley showed that such a cross-bridge cycle could explain many kinetic and thermodynamic aspects of muscle contraction such as the force–velocity curve and the liberation of extra heat during active shortening (the Fenn effect). The 1957 model made a number of testable predictions and stimulated much further research. Many of the assumptions of the model have received experimental support, but some of the later data could not be explained by the 1957 theory, so the model must be enhanced. Several reviews have covered this ground (A. F. Huxley, 1974, 1980; Simmons & Jewell, 1974), so these developments will not be treated here.

One aspect of the 1957 model that is particularly relevant to the experiments discussed in this paper is the postulate that filament sliding may carry an attached cross-bridge into a position where its mechanical strain becomes negative. In the 1957 model, cross-bridges at negative strain support negative force and pose a resistance to shortening. Maximum shortening velocity (V_{max}) is reached at zero load when the forces of positively and negatively strained cross-bridges balance. Detachment of cross-bridges is more rapid at negative strain than at positive strain, which helps to explain that total energy liberation and rate of ATP splitting are higher during shortening than in isometric contractions (Fenn, 1923; Kushmerick & Davies, 1969).

Force–extension curve of active muscle fibres

The elasticity of the cross-bridge has frequently been studied by applying quick length changes to an activated muscle and detecting the concomitant tension response. If the length step is fast enough, the relation between the length change and the tension recorded during the perturbation, termed a T_1 curve, shows the instantaneous force–extension curve of the cross-bridge elasticity. Figure 15.1 qualitatively shows the T_1 curves that are expected if the cross-bridges can support a negative strain (A) or if they buckle under negative load (B). The smaller diagrams at the top show hypothetical T_1 curves for typical cross-bridges. Since the subunit periodicity in the thick filament of vertebrate skeletal muscle does not match that of the thin filament, cross-bridges attached at any moment will have a variety of strains. The distribution of strains is depicted at the top of Fig. 15.1 by a horizontal shift among the cross-bridge T_1 curves. The force produced by the fibre is the sum of the cross-bridge forces (larger panels at the bottom of Fig. 15.1). The total force–extension relation is linear and crosses the abscissa with a steep slope if the cross-bridge force–extension relations are linear (A). If cross-bridges go slack under negative strain, the fibre force–extension relation is curved and crosses the abscissa at a much lower slope (B).

Ford, Huxley, and Simmons (1977) found that the T_1 curve of tetanically stimulated frog single fibres is nearly linear and crosses the abscissa with a steep slope at a length step amplitude of about 4 nm of filament sliding. Tension can become negative briefly during a large quick release. These observations suggest that the cross-bridge force–

Fig. 15.1. Hypothetical force–extension relations for the case that cross-bridges can support negative forces (A) and the case that cross-bridge stiffness decreases at negative strain (B). The distribution of strains among cross-bridges at a given sarcomere length is represented here by horizontal shifts between the five cross-bridge force–extension relations at the top of panels A and B. The bottom shows fibre force–extension relations (T_1 curves) resulting from the summation of cross-bridge forces.

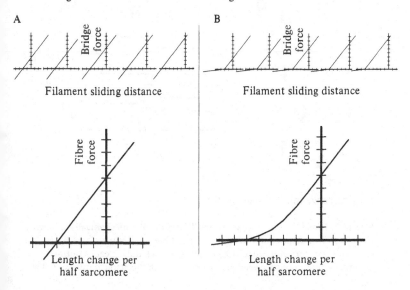

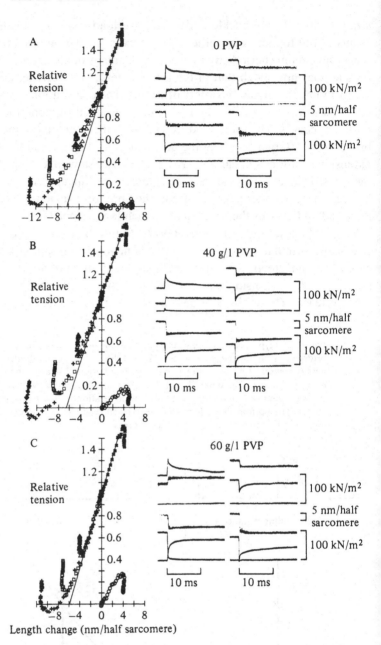

Fig. 15.2. Immediate force–extension (T_1) curves of a mechanically skinned fibre from frog semitendinosus muscle. The fibre was activated at 1.8 μM free Ca^{2+} and transferred to sarcomere length feedback during each contraction. Separate contractions were elicited to record each mechanical transient. The

extension relation is nearly linear and that cross-bridges can support negative forces. After a quick release, total force dwells in the negative range only briefly owing to dynamic recovery processes probably related to the normal power stroke (A. F. Huxley & Simmons, 1971b).

Bob Simmons and I set out to measure in mechanically skinned muscle fibres the sort of mechanical responses Ford, Huxley, and Simmons had investigated in intact fibres. The main purpose was to modify the biochemical conditions within the myofibrils. We sought to avoid problems due to end-compliance in such experiments, and the necessary technical developments are discussed in a later section. Compared to intact fibres, the force–extension relations we observed with Ca^{2+}-activated skinned fibres were noticeably more curved and did not cross the abscissa with a steep slope (Fig. 15.2A). This curvature and lower elasticity in skinned fibres was not due to end-compliance because the sarcomere length step was measured by laser diffraction from a central portion of the fibre segment. A release of more than 12 nm per half sarcomere was required to reduce force nearly to zero (Goldman & Simmons, 1986).

An important difference between skinned and intact fibres is that the filament lattice of a skinned fibre is laterally swollen owing to electrostatic repulsion or Donnan–osmotic forces (Matsubara & Elliott, 1972; Godt & Maughan, 1977). We found that if the filament lattice was osmotically compressed with large synthetic polymers towards the intact fibre dimension (Matsubara, Goldman & Simmons, 1984), the force–extension relation became steeper and less curved (Fig. 15.2B & C; Goldman & Simmons, 1986). An attractive explanation for these results is that the cross-bridges in the swollen filament lattice cannot support negative forces as well as at the normal lateral spacing, so the force–extension relation is curved as in Fig. 15.1B. In intact fibres and in

Caption to Fig. 15.2. *(cont.)* insets show sarcomere length change (noisy traces) and tension response during quick length changes. The tension data, corrected for the transducer response, are plotted against the instantaneous length change corrected for end compliance. The different symbols in each panel show the response for different amplitudes of length change. Superposition of data from different size steps indicates that the response is due to nearly pure elasticity. T_1 curves were obtained at at 0 (A), 40 (B), and 60 (C) g/l polyvinyl pyrrolidone (PVP-40), a linear synthetic polymer that compresses the filament lattice. Points plotted near the origin indicate the resting force–extension relation. Sarcomere length: 2.12 μM. Fibre dimensions (at 40 g/l PVP): 3.17 mm × 11,300 μm^2. T = 3.2°C. [Reproduced from Y. E. Goldman & R. M. Simmons (1986). *Journal of Physiology, London, 378,* 175–94.]

osmotically compressed skinned fibres, the cross-bridges can apparently support a negative force as postulated by A. F. Huxley.

Helix-to-coil transition

Most investigators assume that the structural change in the cross-bridge that tends to cause filament sliding is a tilt of the myosin head. But it had been very difficult to obtain direct evidence for head tilting during the power stroke (Cooke et al., 1984). Harrington and co-workers (Harrington, 1971, 1979; Ueno & Harrington, 1981, 1986; Davis & Harrington, 1987) have postulated instead that the crucial structural change in the cross-bridge is partial "melting" of the α-helical coiled-coil of myosin subfragment-2 (S-2) while the S-1 head is attached to actin. The transition from α-helix to a random coil would decrease the end-to-end distance of the melted section, thereby generating the sliding force.

Harrington's group has obtained extensive data supporting the existence of a helix-to-coil transition. For instance, the susceptibility of S-2 to proteolytic digestion is higher during active contractions than in relaxation or rigor (Ueno & Harrington, 1986). Helix-to-coil melting can explain many of the mechanical properties of active muscle (Harrington, 1979), such as the quick recovery of force following length steps shown in Fig. 15.2.

The shape of the cross-bridge T_1 curve near zero force is relevant to the helix-to-coil hypothesis. Figure 15.3 shows that a nonlinear force-extension curve arises if S-2 has a section of random coil. In the absence of mechanical forces, the random coil has a most likely end-to-end distance (L_i) set by the statistical–mechanical properties of the peptide chain with torsionally flexible bonds at the alpha carbon atoms. When force is applied to the coil, the end-to-end distance increases and the subunit positions become less random. Thus, the elasticity is related to maximisation of entropy as in rubber. The entropic nature of the elasticity leads to an unusual thermoelastic property. Like rubber, a random coil shrinks when heated.

If the force is zero, the entropy is maximal when the end subunit of the melted section is at any point on the hemispherical surface shown as a dashed line in Fig. 15.3A. If a compressive strain is applied, the random coil cannot support the negative force unless it tumbles to the opposite side of the maximum entropy locus as in Fig. 15.3B. Consequently, the cross-bridge force–extension relation (panel C) does not cross the abscissa with a steep slope; instead there is a "dead-band" equal to twice L_i.

Since the intact fibre and the osmotically compressed skinned fibre T_1 curves do cross the abscissa steeply, there seems to be little (< 1 nm) random coil in the cross-bridge during contractions. The thermoelasticity of a fibre during contraction is not rubberlike (Goldman, McCray, & Ranatunga, 1987; Bershitsky & Tsaturyan, 1989; but see Davis & Harrington, 1987), and myosin subfragment-2 is not essential for ATP-driven actomyosin motility in vitro (Toyoshima et al., 1987). These experiments seem to provide strong evidence that helix-to-coil melting is not the primary mechanism for force generation.

The force–extension relations discussed here might help to explain some of the data of Harrington and co-workers. The increased susceptibility of S-2 to proteolysis during contraction has been demonstrated mostly in skinned muscle fibres and myofibrils that presumably have a swollen filament lattice. Taking account of the nonintegral periodicities between the two filaments as in Fig. 15.1B, the hypothetical T_1 curve for a partly random-coil cross-bridge (Fig. 15.3C) would produce a curve much like the skinned fibre force–extension relation (Fig. 15.2A). Perhaps the increased lattice spacing in the skinned fibre enhances the availability of S-2 to the proteolytic enzymes or else more S-2 melts during contractions when the lattice is swollen. It would be useful to determine whether compression of the fibres with osmotic agents alters the rate or specificity of S-2 proteolysis or cross-linking between S-2 and the thick-filament backbone.

Davis and Harrington (1987) have reported that at high temperatures and high pH in rigor fibres, S-2 melts and displays rubberlike thermoelasticity. Can those cross-bridges with random coil zones support negative

Fig. 15.3. Diagrammatic representations of a hypothetical cross-bridge with part of the subfragment-2 melted into a random coil as in the Harrington (1971, 1979) hypothesis. Panel A represents positive strain; B, negative strain. L_i is the most likely end-to-distance of the coil at zero force and maximum entropy. At zero force, the end of the random coil can lie at any point on the hemispherical surface represented by the dashed line. Panel C is the instantaneous force–extension relation of the cross-bridge with some random coil. The slope near zero force is much lower than at high force.

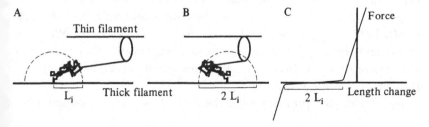

tension? Is their force–extension relation linear? Quick release experiments on rigor fibres at high temperature and high pH would test the prediction of Fig. 15.3 that random coil in the cross-bridge leads to a nonlinear T_1 curve.

Skinned-fibre length clamps

Attachment of muscle fibres into an experimental apparatus is invariably accompanied by compliance in series with the sarcomeres. To monitor or control the filament sliding in mechanical experiments, it is necessary to record either the fibre length away from the end regions or the striation spacing itself. Bob Simmons and I tried the "spot-follower" device that had been used successfully with intact fibres by Ford, Huxley, and Simmons (1977, 1981, 1985, 1986), but with skinned fibres, activation was markedly nonuniform near the metal foil markers used to detect length of the central segment.

We devised the diffraction device shown in Fig. 15.4A to monitor striation spacing. A helium–neon laser beam is directed at the fibre at a meridional angle that can be varied by rotating a mirror mounted on a servo-controlled galvanometer coil. The galvanometer is a smaller version of the motor Lincoln Ford had built to apply quick length changes to fibres (Ford et al., 1977). A set of photodiodes detects the position of the light diffracted by the striations, and the position signal is fed back to the galvanometer coil to keep the diffracted light centred on the photodiode array. Because the system is a null-detecting servo, linearity of the position sensor is not important. The angle of the galvanometer mirror is uniquely related to striation spacing; the latter can be accurately determined at high time resolution by another set of photodiodes. The striation spacing signal from the setup of Fig. 15.4A had electronic noise corresponding to less than 0.04 nm per half sarcomere, DC to 10 kHz.

When we tried to calibrate the instrument against direct microscopic measurements of small changes of striation spacing or when we applied ramp length changes to the fibres, we found the striation spacing signal often changed stepwise rather than smoothly, even in relaxed fibres. The steps were similar to those observed by Pollack and colleagues (Pollack et al., 1977; Delay et al., 1981; Granzier & Pollack, 1985). With our instrument the steps were clearly an artefact, because they correlated with discrepancies between the electronic signal and direct microscopy (Goldman & Simmons, 1984).

We thought the problem might be caused by the Bragg-angle phenomenon described by Rüdel and Zite-Ferenczy (1979, 1980), by analogy into x-ray diffraction of crystal lattices. According to this idea, cross-striations oriented precisely so that the incident and diffracted angles are equal would dominate the intensity of the diffracted light. Then the output signal from our diffraction device would represent the spacing of those striations rather than the average. Andrew Huxley and Michael Ferenczi independently suggested that we might obtain better averaging of the striations if we used the galvanometer mirror to oscillate the laser beam about its steady angle to sample more orientations. That trick eliminated the artefactual steps and reduced the discrepancies between the signal and microscopic estimates to about 0.4 nm per half sarcomere, which is the usable resolution of the instrument (Goldman & Simmons, 1984). But the oscillation method limited the bandwidth of the measurement to below the 3-kHz oscillation frequency that could be used with the galvanometer coil.

When I was preparing to move to Philadelphia and was thinking about what sort of length clamp device to build in my new laboratory, Bob Simmons and Andrew Huxley both suggested that polychromatic light might solve the Bragg-angle problem because each colour would have its own Bragg angle. The next morning Andrew Huxley gave me the diagram shown in Fig. 15.4B. The reproduction here is fuzzy because the photographer had to remove text showing through from the other side of the original sketch, the point being that Andrew Huxley was already recycling paper in 1978.

In the proposed setup of Fig. 15.4B, white light is reflected from a galvanometer mirror as in the laser system. The light is then dispersed into a spectrum by a special doublet lens constructed to have severe chromatic aberration by coupling converging flint glass and diverging crown glass elements. This arrangement is the reverse of a normal achromatic doublet. The various colours are projected onto the muscle fibre by another lens, and a feedback loop positions the diffracted light on the centre of the photodetector. Since the Bragg angle depends on wavelength, the various colours select striations over a range of tilt angles. I never fully understood how Fig. 15.4B was supposed to work, but it did stimulate thought about projecting coloured light onto a fibre at different angles to circumvent the Bragg angle problem.

When I moved to Philadelphia, I built the diffractometer shown in Fig. 15.4C. It uses white light as suggested by Andrew Huxley, but instead of the highly aberrated lens, an acousto-optic (A-O) device disperses the light into a spectrum. A piezo-electric element excites trans-

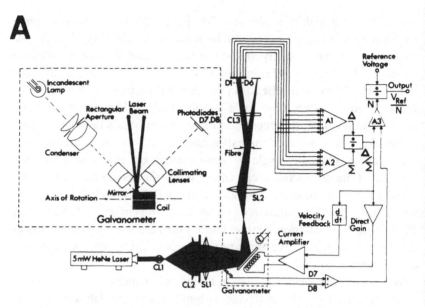

Fig. 15.4. Diffractometers for measurement of muscle fibre striation spacing during mechanical transients. (A) CL1, CL2, CL3: cylindrical lenses; SL1, SL2: spherical lenses; A1, A2, A3: amplifiers; D1–D8: photodiodes; ÷: analog divider circuits; *d/dt:* differentiator. *Inset:* optical arrangement for detection of the mirror angle. The inset in (A) is drawn in the plane perpendicular to that of the main diagram. (B) A sketch drawn by Prof. A. F. Huxley in 1978 proposing a polychromatic diffraction device. A collimated white light beam is refracted into a spectrum by the lower flint/crown doublet lens. Diffraction by the sarcomeres of the muscle fibre combines the various colours into a white light beam (not shown). The photodetector and feedback to the galvanometer mirror are not shown but would correspond to those of panel A. (C) Xe: 75-watt xenon arc lamp; SL1, SL2, SL3, and SL4: achromatic spherical lenses; CL1 and CL2: singlet cylindrical lenses; A-O: acousto-optical light-beam deflector; BS1 and BS2: beam stops; M1 and M2: front surface mirrors (M2 removed for white light excitation); HeNe: helium neon laser; MO: microscope objective; FO: fibre optic light guide. HeNe, MO, FO, SL4, and M2 were used to test the setup with monochromatic light. PD: array of 16 photodiodes. A1, A2, and A3: summing amplifers; ÷: analog divider circuit; DG: direct gain feedback amplifier; *d/dt:* velocity feedback amplifier; V/F: voltage-to-frequency convertor; PA: radiofrequency power amplifier; ÷5120: binary and decade digital counter; BC: binary counter; D/A: digital-to-analog converter. The optical–electronic feedback loop centres the twice-diffracted light onto PD, and the Sarcomere Length Out signal is proportional to striation spacing. [A reproduced from Y. E. Goldman & R. M. Simmons (1984). *Journal of Physiology, London, 350,* 497–518. B reproduced by permission of A. F. Huxley. C reproduced from Y. E. Goldman (1987). *Biophysical Journal, 52,* 57–68, by copyright permission of the Biophysical Society.]

B

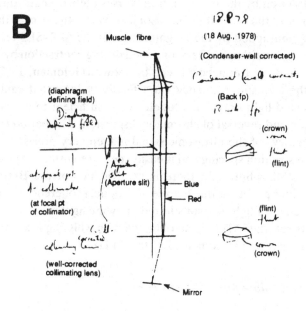

Muscle fibre

⟨handwritten⟩

(18 Aug., 1978)

(Condenser-well corrected)

(diaphragm
defining field)

(Back fp)

(crown)

(flint)

(Aperture slit)

— Blue

— Red

(flint)

(at focal pt
of collimator)

(crown)

(well-corrected
collimating lens)

Mirror

C

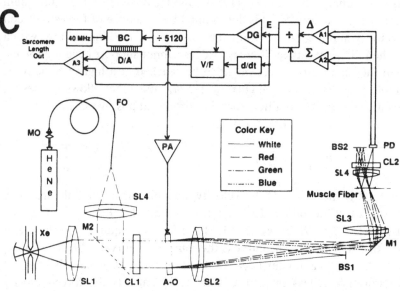

verse ultrasonic acoustic waves at 30–75 MHz across the optical medium
of the A-O device. The periodic changes of refractive index due to the
ultrasonic waves scatter light into the spectrum that is projected onto the
fibre by a telescope composed of lenses SL2 and SL3 (Fig. 15.4C). A

second diffraction by the muscle fibre reverses the chromatic dispersion and recollimates the beam if the ultrasonic wavelength matches the striation spacing multiplied by the magnification of the SL2–SL3 telescope. This matching condition can be maintained during contraction by varying the ultrasonic drive frequency. It can be shown (Goldman, 1987b) that adjusting the acoustic frequency via a feedback loop that centres the twice-diffracted light on the photodetector is sufficient to maintain the recollimation and reversal of chromatic dispersion. The period (reciprocal of frequency) of the ultrasonic signal is then very precisely proportional to the striation spacing and is easily measured using a digital counter circuit. With white light there is no evidence of the Bragg-angle artefact or other problems due to excessive optical coherence (Goldman, 1987b). When laser light was substituted for white light in the same setup, the artefacts returned. The signal obtained with white light was useful to about 0.1 nm per half sarcomere, DC to 10 kHz.

Skinned-fibre force–velocity curve

As mentioned, the Huxley (1957a) model postulates that maximum shortening velocity (V_{max}) is determined by a balance of forces between positively and negatively strained cross-bridges. In skinned fibres with a swollen filament lattice, if cross-bridges cannot support negative forces, then the balance of forces ought to be reached at a higher velocity. A simple modification to the Huxley (1957a) model that allows stiffness at negative strain (k_2) to differ from the stiffness at positive strain (k_1) leads to the following equation for the force–velocity curve:

$$P = P_o \cdot \left\{ 1 - \frac{V}{\Phi}(1 - e^{-\Phi/V}) \cdot \left[1 + \frac{1}{2} \cdot \frac{k_2}{k_1} \cdot \frac{(f_1 + g_1)^2}{(g_2)^2} \cdot \frac{V}{\Phi} \right] \right\}$$

where P is the force, V the velocity, and the other symbols are as in Huxley's (1957a) paper. This equation is plotted in Fig. 15.5A for various ratios k_2/k_1. When $k_2/k_1 = 1$, the stiffness is equal for positive and negative strains, and the curve is the same as in Huxley (1957a). If $k_2/k_1 < 1$, then V_{max} is higher because the drag force of negatively strained cross-bridges is less than at $k_2/k_1 = 1$. k_2/k_1 affects the force–velocity relation mainly in the low force region.

These points were tested on mechanically skinned frog fibres using the white light diffractometer to measure sarcomere shortening velocity. Ca^{2+}-activated skinned fibres shortened very rapidly (Fig. 15.5B). Shrink-

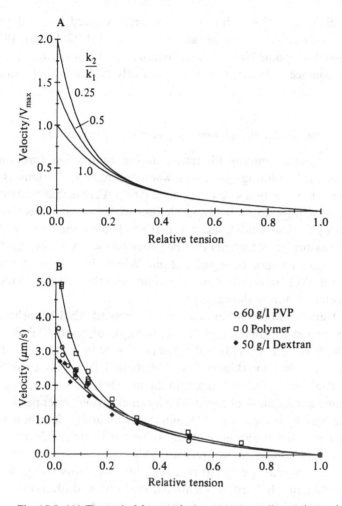

Fig. 15.5. (A) Theoretical force–velocity curves according to the model of Huxley (1957a) modified to incorporate reduced cross-bridge stiffness at negative strain. k_2/k_1 is the ratio of cross-bridge stiffness at negative strain (k_2) to that at positive strain (k_1). Force and velocity are plotted relative to the isometric tension and V_{max} corresponding to $k_2/k_1 = 1$. As k_2 is reduced, the predicted velocity at low loads increases. (B) Experimental force–velocity curves of a mechanically skinned frog fibre. Velocities measured by white light diffraction 10 ms after the load step are plotted against isotonic tension relative to the isometric tension in each solution. Polymers polyvinyl pyrrolidone (PVP-40) and dextran (T-500) reduce shortening velocity mainly at low loads. [Ca^{2+}] = 0.98 μM. Fibre dimensions, 3.18 mm × 31,800 μm^2. Sarcomere length = 2.32 μm during the isometric phase. T = 3.5°C. [Reproduced from Y. E. Goldman (1987). *Biophysical Journal 52*, 57–68, by copyright permission of the Biophysical Society.]

ing the filament lattice with osmotic polymers reduced V_{max} and altered the force–velocity curve mainly at the low force end (Goldman, 1987b). These results support Huxley's idea that the zero net force at V_{max} represents a balance between forces of positively and negatively strained cross-bridges.

Cross-bridge detachment by photolysis of caged ATP

Since the rate of energy liberation during contraction (presumably linked to ATP splitting) increases when the muscle shortens (Fenn, 1923), the slow reaction step of the actomyosin ATPase that controls the isometric cycling rate probably accelerates when the filaments slide. In the Huxley (1957a) model, the step that accelerates when a fibre shortens is cross-bridge detachment; it becomes fast when a cross-bridge is brought into the region of negative strain. Which biochemical step in the actomyosin ATPase is rate-limiting in isometric contractions? Does that step accelerate during shortening?

Jody Dantzig, Michael Ferenczi, Mark Hibberd, Earl Homsher, Jim McCray, David Trentham, and I used photolysis of caged ATP (Kaplan, Forbush, & Hoffman, 1978) and caged phosphate to investigate rates of the biochemical steps in skinned fibres (Hibberd & Trentham, 1986; Goldman, 1987a). ATP is liberated within the muscle fibre approximately 10 ms following irradiation of caged ATP by an ultraviolet laser pulse. With the fibre initially in rigor, the ATP initiates cross-bridge detachment and either relaxation in the absence of Ca^{2+} or reattachment and development of steady active tension in the presence of Ca^{2+}. Using skinned fibres from rabbit psoas muscle, we found that ATP binding, cross-bridge detachment (Goldman, Hibberd, & Trentham, 1984a,b), hydrolysis to protein-bound ADP and P_i (Ferenczi, Homsher, & Trentham, 1984; Ferenczi, 1986), cross-bridge reattachment and P_i release (Hibberd et al., 1985; Webb et al., 1986; Dantzig et al., 1987; Dantzig & Goldman, 1989) are all much faster than the 300- to 400-ms turnover time of the ATPase reaction during an isometric contraction. The power stroke is linked to P_i release from an actomyosin (AM)–ADP–P_i state. Rates of some of these steps seem to be strain dependent (Webb et al., 1986; Homsher & Lacktis, 1988) but since they are fast, they probably do not appreciably control the rate of overall turnover.

Relaxation following photolysis of caged ATP in the absence of Ca^{2+} is markedly slowed by adding 50 μM or more MgADP to the medium (Dantzig et al., 1991). Although the simplest explanation for this result might be that ADP release from AM–ADP is slow, the situation is more complicated. Even though Ca^{2+} is absent, some cross-bridges reattach and generate force because the thin filaments are turned on by other attachments (Goldman et al., 1982; 1984a). The reattached cross-bridges contribute to the pronounced slowing of relaxation when MgADP is present. The time course of relaxation in the presence of MgADP is complex: a plateau phase of steady or rising tension is followed by a sudden transition to more rapid relaxation. Applying a small stretch either before the photolysis laser pulse or during the plateau initiates the rapid relaxation phase, indicating that the net cross-bridge detachment rate is strain dependent. The direction of this strain dependence is opposite to that expected from the Huxley (1957a) model. However, a better comparison with the model is obtained when negative strain is applied to the cross-bridges.

Figure 15.6 shows a method to generate negative strain in rigor cross-bridges (Goldman, McCray, & Vallettee, 1987). A skinned single fibre from rabbit psoas muscle is stretched in relaxing solution to generate significant resting tension (left panel, point A) above zero (thin dashed line). It is then put into rigor and washed again with rigor solution to remove residual ATP. A small length release decreases tension to below the relaxed level (point C). Thus, at long sarcomere lengths, the rigor force–extension curve extends below the relaxed one, indicating a region of negative cross-bridge stress and strain (Fig. 15.6, right panel). When MgATP is added in the absence of Ca^{2+} (labelled *Relax* in Fig. 15.6, left), tension increases and stiffness decreases (not shown) as cross-bridges bearing negative force detach.

Relaxation of positive and negative tension by photolysis of caged ATP is shown in Fig. 15.7. In the absence of ADP, relaxation is rapid from either positive or negative tension (A). When 50 μm to 1 mM MgADP is added, relaxation from positive tension (i) is slowed (B, C). Relaxation from negative tension (r) is less affected by MgADP, but at 2 mM MgADP, negative relaxation is also slowed (panel D; Dantzig et al., 1991). Since ADP dissociation precedes detachment, these results suggest that at negative strain, ADP dissociates faster from AM–ADP than at positive strain, perhaps due to a strain-dependent isomerisation of AM–ADP. During the cross-bridge cycle in a shortening muscle, if ADP release is slow until the cross-bridge reaches negative strain, the detachment by ATP would show just the sort of strain dependence postulated by Huxley. However, we are not certain that the initial rigor AM and AM–ADP states in the caged ATP experiment are on the normal reaction pathway during a contraction. Also, there is no Ca^{2+} present in the experiment. Thus, the applicability of the data of Fig. 15.7 to a normal contraction is indirect. Furthermore, not only slow ADP release but also cross-bridge reattachment contribute to the delay of relaxation in Fig. 15.7A–D. Nevertheless, the results suggest a strain dependence of net detachment in the direction proposed in the 1957 theory: negatively strained cross-bridges detach more rapidly.

Fig. 15.6. Procedure to apply negative strain to cross-bridges in rigor. *Left panel:* A relaxed glycerol-extracted single fibre from rabbit psoas muscle was stretched to generate significant resting tension (point A) above zero (thin dashed line). It was then put into rigor solution [two washes (arrows) to remove residual ATP; point B]. A small length release (120 μm) decreased tension to below the relaxed level (point C). Addition of MgATP (*Relax*, final concentration 0.7 mM) then caused tension to *increase*. At *Stretch*, the fibre length was increased to the prerigor length; and at *Slacken*, the zero force level was recorded. Fibre dimensions: 2.28 mm × 4800 μm², measured at a sarcomere length of 2.40 μm. The recording starts at a sarcomere length of 3.35 μm. *T* = 20°C. *Right panel:* force–extension curves from another fibre. A, B, and C indicate corresponding conditions in the two fibres. [Reproduced from Y. E. Goldman, J. A. McCray, & D. P. Vallette (1987). *Journal of Physiology, London, 398*, 72 P.]

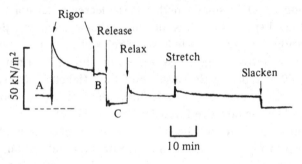

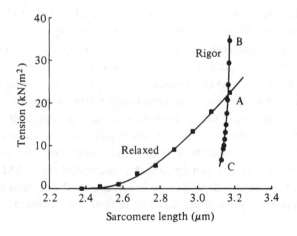

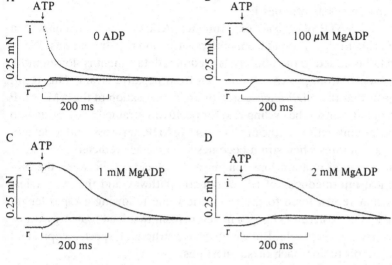

Fig. 15.7. Effect of MgADP on the relaxation kinetics of positively and negatively strained cross-bridges in the absence of Ca^{2+}. The preparation was a
glycerol-extracted single fibre from rabbit psoas muscle. The fibre was relaxed by photo-release of 1 mM ATP from caged ATP in rigor solution with
the indicated MgADP concentrations. In each panel, the top trace (i) is an
isometric trial at positive tension; sarcomere length = 3.3 μm. The centre
trace in each panel shows the tension of the relaxed fibre recorded about 1
min later. The fibre was then put into rigor again and released by 3% to
apply negative cross-bridge strain. The lowest trace (r) show relaxation of
negative cross-bridge tension on photo-release of ATP. For clarity the negative tension records (r) and their baselines were shifted vertically to superimpose all of the baselines. Fibre dimensions: 1.99 mm × 8200 μm^2, measured
at a sarcomere length of 2.15 μm. T = 20°C. [Reproduced from J. A.
Dantzig et al. (1991). *Journal of Physiology, London*, 432, 639–680.]

Conclusions

Results from probing the mechanics and kinetics of cross-bridges under
negative strain are consistent with the ideas first articulated by Andrew
Huxley in his 1957 model. In skinned muscle fibres with a laterally
swollen filament lattice, the force–extension relation is curved and does
not cross the abscissa at steep slope. Active shortening velocity is high as
expected if cross-bridges cannot support negative strain in the swollen
lattice, and therefore, shortening is less restrained than in intact fibres.
The velocity and force–extension effects are both reduced by compressing the filament lattice with high-molecular-weight polymers. These results suggest that the swelling is the cause of the curved force–extension

relation and the elevated velocity, and that a balance of positive and negative forces determines V_{max}.

In caged-ATP photolysis experiments, ATP-induced detachment of rigor cross-bridges is rapid from both positive and negative strains. When MgADP is added to the photolysis medium, detachment is slowed more prominently from positive strain than from negative strain. This result suggests that the ADP release step or an isomerisation of the AM–ADP state might control the cycling rate for positively strained cross-bridges in a normal contraction. In the presence of MgADP, net cross-bridge detachment accelerates when strain becomes negative as predicted by Huxley.

The detailed relationships remain to be elucidated between the biochemical intermediates of the enzymatic pathway and the mechanical steps that return force to the isometric value following a rapid length change. The structural changes in actomyosin that lead to energy transduction are still a puzzle. But we now have structural, spectroscopic, and kinetic tools to approach these questions.

16

Force response in steady lengthening of active single muscle fibres

VINCENZO LOMBARDI AND
GABRIELLA PIAZZESI

Introduction (by Vincenzo Lombardi)

My collaboration with Professor Huxley has been the most important event of my scientific career. My work with him enabled me to investigate the mechanical aspects of contraction in intact fibres from frog skeletal muscle at a level very close to the functional unit, the half sarcomere. During the time Lee Peachey and I spent in his laboratory at University College, London (October 1979 to July 1980), Professor Huxley developed the striation follower. This is an optoelectronic device that measures, with a time constant of 1 μs, the change in length undergone by a short segment of a contracting striated fibre. I must confess that I was able to build a copy of the striation follower when I was back in Florence only as a result of the enthusiasm and perseverance of Professor Huxley in trying to involve his co-workers in the process of getting the best solutions to the problems he is facing during the realisation of an idea.

My succeeding research activity has simply been the most obvious consequence of my collaboration with Professor Huxley. I have been able to do reliable experiments on an intriguing aspect of contraction, namely length control at sarcomere level. Most of the ideas and developments inspired by the experiments I describe in this chapter received invaluable support from Professor Huxley. I am grateful for the criticisms and suggestions he has given me throughout the decade that has passed since my first experience of working with him. During this period I was able to appreciate one of his more unusual characteristics: his generous enthusiasm for the reliable experimental results of others – an enthusiasm most scientists demonstrate only towards their own results.

A. F. Huxley's theory of muscle contraction

It is generally accepted that the contractile properties of striated muscle result from the cyclic interaction between the projections from the myosin filament (the cross-bridges) and specialised sites on the actin filament (H. E. Huxley, 1953; H. E. Huxley & Hanson, 1954; A. F. Huxley & Niedergerke, 1954). The dynamics of cross-bridge interaction can be characterised on the basis of the mechanical and thermodynamic properties of active muscle. It has long been known that the velocity of shortening of a stimulated muscle increases with the reduction of the applied load below the isometric tetanic force (T_0), in accordance with the empirical force–velocity relation of Hill (1938), as shown in Fig. 16.2, and that the total energy liberation, which is proportional to the extent of the chemical reactions coupled to contractile activity, is higher when the muscle shortens than under isometric conditions (Fenn, 1923, 1924).

In 1957, A. F. Huxley formulated a model of contraction to explain the relation among force, energy liberation, and velocity of shortening. In the model, the cross-bridge can exist in two states – attached and detached. The force of an attached cross-bridge depends on the degree of extension of an elastic link between the cross-bridge and the actin site to which it is attached (A. F. Huxley, 1957a). The capability of developing force or shortening depends on a limited number of parameters, such as the number of actin sites available for interaction with the cross-bridges and the values that the rate constants for cross-bridge attachment (f) and detachment (g) assume as function of x [where x, as defined in Huxley's (1957a) paper, is the displacement of a cross-bridge from its equilibrium position]. The value of f is moderate but larger than g. During shortening, cross-bridges are shifted to negative values of x, where f is zero and g becomes large. Consequently, the number of attached cross-bridges decreases and the turnover rate increases with an increase in shortening velocity. Actually, in agreement with the model, stiffness (which in the absence of a passive series compliance represents an estimation of the number of attached cross-bridges) has been found to decrease (Julian & Sollins, 1975; Ford, Huxley, & Simmons, 1985) and ATP hydrolysis to increase (Kushmerick & Davies, 1969) during steady shortening.

Force during lengthening

The contractile response to lengthening represents an interesting condition for further investigating the characteristics of cross-bridge interaction. There are a great number of intriguing results and ideas about the events occurring during stretch. For instance, in spite of the high level of force attained during and after lengthening, the energy liberation is much lower than in isometric conditions (Fenn, 1924; Hill, 1938; Aubert, 1944a,b, 1948).

 The experiments referred here were aimed at investigating the mechanical aspects of the response to lengthening imposed on active muscle fibres under conditions in which the actual length change of a selected population of sarcomeres was measured, either in fixed-end or in length-clamp mode, by means of a striation follower (A. F. Huxley, Lombardi, & Peachey, 1981). This instrument measures the longitudinal displacement of two distinct regions along a muscle fibre by counting the number of sarcomeres crossing the optical field of a two-channel optoelectronic device. The difference between the ouput signals from the two channels gives the actual length change of the segment (1–2 mm long) delimited by the two regions. Experiments were performed at low temperature (2–5°C) and at a sarcomere length of about 2.05 μm, on single fibres (4.5–5.5 mm long) dissected from the lateral head of the tibialis anterior muscle of the frog (*Rana esculenta*). Ramp lengthening (velocity between 0.02 and 1.2 μm/s per half-sarcomere, h.s.) not larger than 5% of the fibre length was imposed on the tetanised fibres so that, at the end of the ramp, the sarcomere length never exceeded 2.2 μm. For the characteristics of the force transducer and the loudspeaker motor, see A. F. Huxley and Lombardi (1980), Ambrogi Lorenzini, Colomo, and Lombardi (1983) and Cecchi et al. (1987).

 When a steady lengthening is applied at the plateau of an isometric tetanus, force increases above the plateau and then, beyond a critical amount of lengthening (10–13 nm per h.s.), attains a roughly steady value (Fig. 16.1). At velocities higher than 0.3 μm/s per h.s., force rises to a peak before declining to the steady value. There is a great deal of variability in the literature on the time course of the force response as well as on the final level of force attained (Edman, Elzinga, & Noble,

1978; Flitney & Hirst, 1978; Julian & Morgan, 1979b), but more recent experiments (Colomo, Lombardi & Piazzesi, 1988) have shown that this variability is due largely to the failure to obtain homogeneous lengthening of different populations of sarcomeres along the muscle fibres, and that it increases with lengthening speed. In the experiments described here, fibres were selected for a high degree of homogeneity: fibres were discarded when, for lengthening at high velocity, segment speed differed more than 30% from the speed imposed by the motor. The actual lengthening occurring at the level of a segment selected along the fibre was continuously monitored by means of the striation follower.

In agreement with the results obtained by Katz (1939) using isotonic lengthening, there is a discontinuity in the relation between steady force and lengthening velocity across the isometric point (Fig. 16.2): the rise in force caused by a slow lengthening is much larger than the reduction in force produced by the same velocity of shortening. Starting from velocities of about 0.3–0.4 μm/s per h.s., an increase in lengthening velocity produces a progressively lower increase in tension, so that with high-speed lengthening, tension attains a limiting value of about 2 T_0.

Fig. 16.1. Tension responses to steady lengthening imposed at the tetanus plateau under segment length-clamp conditions. Figures close to the records indicate the lengthening velocity of the segment (μm/s per h.s.). In each frame, the upper trace is segment length, the lower trace tension response. The switch from fixed-end to length-clamp conditions is marked by the small bars.

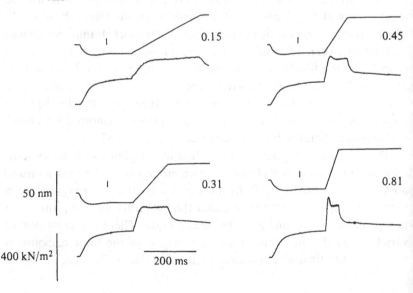

Both these properties of active muscle (the development of high forces through very slow lengthening and the attainment of a maximum force through fast lengthening) can be explained in terms of Huxley's 1957 model: with slow lengthening, the average strain of attached cross-bridges rises while further attachment takes place, contributing to the enhancement in force; as the velocity of stretch increases, force attains a limit because of the balance between the increase in average cross-bridge strain and the reduction in their number, due to an increased detachment rate.

Stiffness during lengthening

Information about the nature of the force response to lengthening can be provided by measuring fibre stiffness during steady lengthening at different velocities. Stiffness was estimated by determining the instantaneous tension change produced by small stepwise length changes (<1.5 nm per h.s., complete in about 100 μs) imposed on one end of the fibre. The contribution of the series compliance due to tendon attachments was eliminated by calculating the stiffness as the ratio between the tension change produced by the step and the actual length change of a fibre

Fig. 16.2. Relation of steady-state force (relative to the value at the isometric tetanus plateau) against the velocity of shortening or lengthening. Velocity is plotted on the ordiante. (Lengthening is negative in accordance with Hill, 1938; & Katz, 1939.)

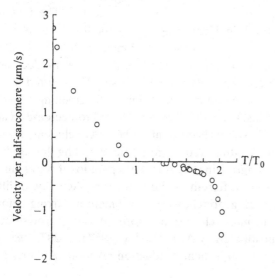

segment, measured by the striation follower. Under these conditions, most of the compliance originates from the cross-bridges themselves (Ford, Huxley, & Simmons, 1981), and, if one assumes that the cross-bridge elasticity is the same during lengthening as in isometric conditions, then changes in stiffness indicate corresponding changes in the number of attached cross-bridges.

As shown in Fig. 16.3, stiffness rises quickly as the lengthening starts and attains an approximately steady value (10–20% higher than at the isometric tetanus plateau) before the force (Colomo et al., 1988). A quantitative estimate of the stiffness increase depends on the procedure used to compare stiffness during lengthening with that determined at the isometric tetanus plateau (Lombardi & Piazzesi, 1990). Stiffness values plotted in Fig. 16.3 were obtained by normalising the stiffness determined at a given force level during lengthening relative to the slope of the isometric tension–extension relation in the same region of force. With this procedure the increase in stiffness during lengthening is about 12%. In any case, regardless of the comparison procedure, these experiments show clearly that during steady lengthening, (1) the increase in stiffness is much lower than the increase in force, so that the ratio of relative force to relative stiffness attains a maximum value of about 1.85; (2) within the range of velocities used, stiffness remains practically constant, independent of lengthening velocity (Fig. 16.3C).

An increased force/stiffness ratio strongly suggests that the force enhancement during lengthening of active muscle is due to an increase in the average extension of attached cross-bridges, with a minor contribution due to a small increase in the number attached. The finding that stiffness is not reduced significantly as the lengthening speed is increased up to 1.2 μm/s per h.s. indicates that attachment in this case is relatively fast, so that the number of attached cross-bridges does not decrease as a result of an increased rate of detachment. It should be noted that, with shortening speeds comparable to the highest lengthening speed used here (1.2 μm/s per h.s.), stiffness has been found to decrease to at least 60% of the isometric value (Ford et al., 1985). Attachment, therefore, appears to be a rate-limiting step in cross-bridge cycling during shortening but not during lengthening. This is not a paradox if we assume that cross-bridge cycles are different in the two cases. Actually, a different cross-bridge cycle during stretch, with detachment occurring before the completion of the normal cycle and therefore without hydrolysis of ATP, has already been postulated by A. F. Huxley (1957a, p. 292; 1980, p. 94) to explain the depression in heat production (Abbott, Aubert, & Hill,

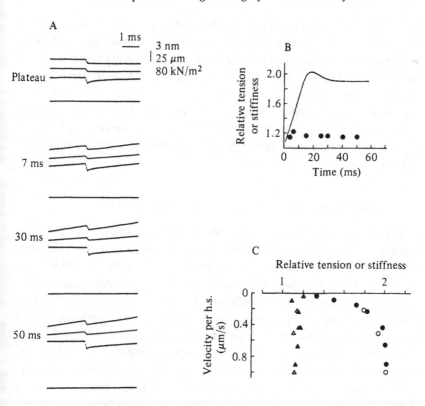

Fig. 16.3. (A) Sample records of tension responses to a small step release (1.06 nm per h.s.) applied to a fibre during the isometric tetanus plateau and at different times (indicated close to each record) after the beginning of the steady lengthening. In each frame, traces from top to bottom show change in segment length, change in fibre length, tension response, resting tension. Same fibre as in Fig. 16.2. Segment lengthening velocity: 0.72 μm/s per h.s. (B) Same fibre segment as in (A). Time courses of force (continuous line) and stiffness (circles) during lengthening at 0.72 μm/s per h.s. Lengthening began at zero time and ended at the interruption of the force record. (C) Relation of the steady-state force (circles) and stiffness (triangles) versus lengthening velocity. Open symbols refer to the mean value for force and stiffness obtained in seven segments of as many fibres. Vertical and horizontal S.E.s are within the size of symbols. In (B) and (C), tension values are expressed relative to T_0, whereas stiffness values are normalised for the non-linearity of the T_1 relation under isometric conditions as described in the text. Temperature: 4.68 ± 0.28°C (mean & S.E.). Filled symbols refer to data from an individual fibre.

1951) and ATP utilisation (Infante, Klaupiks, & Davies, 1964; Curtin & Davies, 1973). A different mechanism of detachment during lengthening is supported here on mechanical grounds by the finding that the degree of strain of attached cross-bridges is much higher than under isometric conditions, as indicated by the larger stiffness/tension ratio. Moreover, the finding that at high velocities force attains a limiting value in the absence of any reduction in stiffness indicates that the rate constant for cross-bridge detachment becomes very high beyond a critical amount of strain.

Tension transients: events at cross-bridge level

Experimental work showing that the shortening velocity of the contractile machinery is not instantaneously determined by the load on it (velocity transients; see Podolsky, 1960; Civan & Podolsky, 1966; Armstrong, Huxley, & Julian, 1966) and that the rate of isometric tension recovery after a stepwise perturbation in length can be resolved in several phases (tension transients; see A. F. Huxley & Simmons, 1971a) was the basis for Huxley and Simmons's (1971b) hypothesis about the mechanism of force generation. The attached cross-bridge exists in different configurations or states, which under isometric conditions exert different forces; once attached, the transition of a cross-bridge towards a state generating more force occurs following a gradient of chemical energy, while work is done on the elastic component of the cross-bridge. Since this work is a component of activation energy for the transition, the rate of the transition and the equilibrium between states will depend also on the mechanical conditions – that is, cross-bridge extension – under which the transition occurs. According to the model, the phases of the tension transient following a stepwise length change (Fig. 16.4), imposed in the absence of a significant passive compliance in series, represent specific events of cross-bridge activity:

Phase 1, the simultaneous tension change produced by the step, is due to the change in length of the elastic component of each cross-bridge. The curve relating the extreme tension attained during the step (T_1) to the step size describes the tension–extension relation of the cross-bridge elastic component (filled circles in Fig. 16.6).

Phase 2, the early tension recovery complete in 1–2 ms, is due to the redistribution of cross-bridges among different configurations caused by the change in cross-bridge extension, without substantial detachment and reattachment. Actually, the rate of early recovery increases continuously from the largest stretch to the largest release (A. F. Huxley & Simmons, 1971b and Fig. 16.6B, circles). The curve relating the tension attained during early recovery (T_2) to the step size (open circles in Fig. 16.6) is flat for relatively small releases; it then declines and attains a constant slope, intercepting the length axis for releases not larger than 12 nm per h.s.

Phase 3, the successive reduction or even reversal of rate of tension recovery lasting a few ms, has a less straightforward explanation. Phase 3 following step releases has been attributed mainly to the lag of the formation of new attachments with respect to the increased rate of detachment that occurs as a consequence of the shift of cross-bridges along the x axis (Huxley & Simmons, 1973).

Phase 4, the final recovery of tension to the value before the step, is due to the detachment of cross-bridges and their reattachment further along the thin filament, until their number and force distribution gradually regain the isometric plateau values (A. F. Huxley & Simmons, 1973).

The modifications undergone by the tension transients elicited during steady shortening (Ford et al., 1985) represent a striking test for the idea that the various phases reflect events at the level of the attached cross-bridge and for the consequent assumptions in terms of cross-bridge kinetics. During shortening, the intercepts of T_1 and T_2 curves on the length

Fig. 16.4. Record showing the sequence of events during the tension transient following a step release (110 μs duration) imposed on a fibre at the isometric tetanus plateau under fixed-end conditions. Traces from top to bottom show tension, segment length, and fibre length. The small bars close to the tension record indicate a 100-fold change in the time base. Numbers refer to the various phases of the tension transient. T_2 (the tension reached at the end of phase 2) was measured following the procedure of Ford et al. (1977).

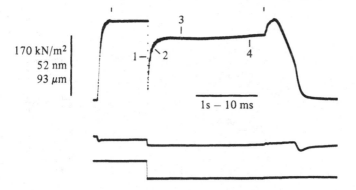

170 kN/m² 52 nm 93 µm

1s – 10 ms

axis are shifted to the right with respect to the isometric ones, indicating a lower extension of attached cross-bridges and a consequent redistribution towards states further ahead in the power stroke. Moreover, as expected by the lower degree of cross-bridge extension, the rate of early recovery for a given step size is higher during shortening than under isometric conditions.

Tension transients during slow lengthening

The study of the tension transients elicited during steady lengthening seems very interesting, not only for testing the possibility of extending the predictions of Huxley and Simmons's (1971b) model to a very critical point, but also for further investigating the dynamics of the cross-bridge cycle. In Fig. 16.5, the tension transients elicited during the steady phase of force response to lengthening at 0.087 μm/s per h.s. are compared with those at the isometric tetanus plateau. As expected from stiffness measurements, the tension change simultaneous with the step is always larger during lengthening than in the isometric state. The tension recovered at any time during the early phase is smaller during lengthening for small stretches and releases, but it becomes larger for large releases; there is no clear reversal or pause in the rate of tension recovery with respect to the rate of final recovery; the time required for the reattain-

Fig. 16.5. Tension transients after length steps imposed in fixed-end mode at the isometric tetanus plateau (left column) and during slow lengthening (middle column). In each row, the figure between the records indicates the approximate size of the segment length change (nm per h.s.) at the end of phase 2. The time base between the small bars close to the tension records is 100 times faster than beyond them. In the right column, the early parts of the same transients are superposed on a faster time base. The extreme tension reached during steps was taken as the reference point for superimposition.

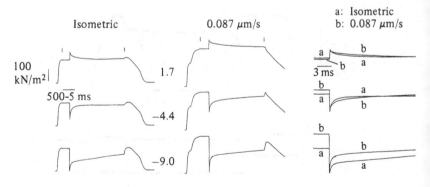

ment of the steady-state tension is shorter. The increase in speed of recovery during this phase is more marked for step stretches than for step releases. Apart from responses to step stretches, phases 2 and 4 are still easily distinguishable. A quantititive estimation of the level to which tension recovers during phase 2 (T_2) can be obtained for step releases by the intercept of the tangent, fitted to the late slower part of tension recovery, with the tension trace at the time of the step (Ford et al., 1977, 1985).

T_1 and T_2 curves determined during slow lengthening are plotted together with those determined under isometric conditions in Fig. 16.6A. In accordance with previous work (Ford et al., 1977), in the isometric state as well as during steady lengthening, the T_1 curve deviates slightly from linearity, showing an upwards concavity that increases with the size of release. The straight line drawn through T_1 points obtained with small releases (which can be considered representative of the properties of cross-bridge elastic component) is steeper during lengthening than under isometric conditions; thus, while force before the step is more than 60% larger during lengthening, the intercept on the length axis is shifted

Fig. 16.6. Same fibre as in Fig. 16.5. (A) T_1 (filled symbols) and T_2 (open symbols) relations under isometric conditions (circles) and during slow lengthening (triangles). T_1 and T_2 values were plotted relative to tension at the isometric tetanus plateau. (B) Rate constant (r) of phase 2 against releases. Symbols as in (A).

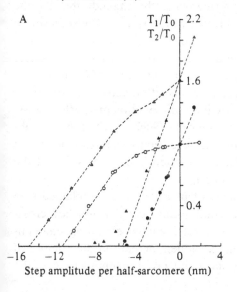

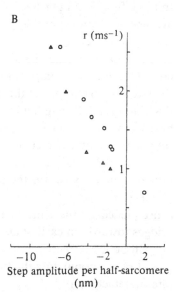

to the left by only about 1.5 nm per h.s. – that is, by about 40% with respect to the isometric value (4 nm per h.s.). The T_2 curve during lengthening is steeper in the region of small releases; it exhibits an inflexion for releases of about 4–6 nm per h.s. and a roughly constant slope for larger releases, intercepting the length axis at about 15 nm per h.s., a value 3 nm per h.s. to the left of the intercept under isometric conditions.

The speed of early tension recovery in transients following step releases of small and moderate size can be estimated by assuming, for the sake of simplicity, that the time course of recovery is described by a single exponential. r, the reciprocal of the time necessary to attain 63% of the early recovery $(T_2 - T_1)$, is plotted against step size in Fig. 16.6B. It can be seen that, for any given step size, r is lower during lengthening than in isometric conditions, so that the whole relation is shifted to the left.

Huxley and Simmons's (1971b) theory provides a straightforward interpretation of the modifications induced by slow lengthening on the tension transient. The reduction of amount and speed of phase 2 for step releases of small and moderate size is an expected consequence of the increased cross-bridge extension at the time of the step. If we assume that, for large releases, rapid detachment of negatively strained cross-bridges does not contribute substantially to early recovery, the intercept of the T_2 curve on the length axis indicates the maximum amount of release for which cross-bridges can recover force while they are attached. A shift of the intercept to the left by 3 nm per h.s. during steady lengthening indicates that the average position (x) of the cross-bridge, with respect to the actin site to which it is attached, is shifted towards more positive values by the same amount. On the other hand, the average increase in the extension of cross-bridge elastic component is only 1.5 nm, as indicated by the shift to the left of the intercept of the T_1 curve. This difference strongly suggests that during steady lengthening there is a redistribution of attached cross-bridges towards a low force generating state. Such a shift of the equilibrium between attached cross-bridges is expected as a consequence of the reduction, with the increased degree of extension, of the value of the forward rate constant for the transition.

The reduction and disappearance of phase 3 can be explained as a consequence of the shift in the equilibrium between attached cross-bridges towards an early stage. A step release imposed during lengthening will be less effective in causing a sudden increase in the number of cross-bridges populating the final stage and a consequent increase of the rate of detachment.

An interpretation of the increase in speed of phase 4 will be given in the following section on the basis of the results obtained with faster lengthening.

Tension transients during fast lengthening

When the speed of lengthening is increased, the extent of phase 2 further decreases while the rate of recovery during phase 4 progressively increases (Fig. 16.7). Consequently, the contribution of phase 2 to the whole recovery is reduced while, at the same time, a quantitative estimate of T_2 by tilting a tangent becomes more and more arbitrary. As shown in Fig. 16.8A, with lengthening at high speed (>0.5 μm/s per h.s.), the time for tension to recover 95% of the value before the step is

Fig. 16.7. Tension response to a small step release (1.06 nm per h.s.) applied during the steady force response to lengthening at different speeds (b) superposed on the response to the same step applied at the isometric tetanus plateau (a). The extreme tension reached during the step was taken as the reference point for superposition. Figures close to the records indicate the lengthening speed. The tension at the isometric tetanus plateau was 180 kN/m². Note that the difference in the instantaneous tension drop between the two superposed records is fairly constant throughout the whole range of velocities, indicating a constant difference in stiffness independent of lengthening speed. Same fibre as in Fig. 16.2.

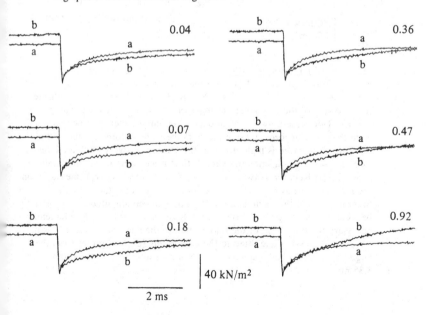

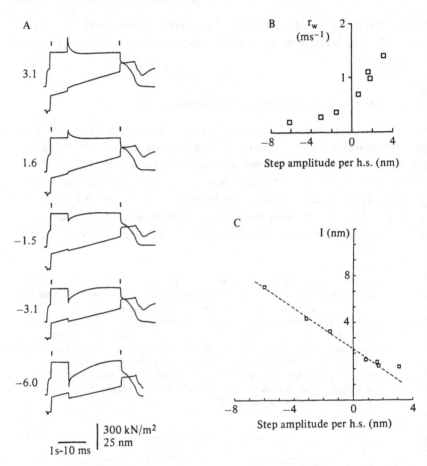

Fig. 16.8. (A) Tension recovery after length steps imposed during fast lengthening (0.7 μm/s per h.s.). In each frame the upper trace is tension and the lower trace is segment length. Numbers to the left of each record indicate size and sign of the step length changes of the fibre segment, in nanometers per h.s. Time base between small bars is 100 times faster than beyond them. Note that in all records, tension recovers the value before the step within the time the trace speed is high, and that the tension recovery after the largest step releases shows a small but very distinct rapid component comparable to phase 2. (B) Relation between the rate of whole recovery (r_w), measured on the same records as in (A) by the reciprocal of the time taken by the tension to attain 63% of the value before the step, and step amplitude. (C) Relation between lengthening (*l*) and step size, where *l* is the amount of lengthening necessary to attain 95% of the tension before the step. Same records as in (A). The regression line fitted to the data points (excluding that from the large step stretch) had a slope of −0.65 and an intercept on the ordinate of 3.35 nm.

within 10–15 ms for step releases and 2–4 ms for step stretches – that is, respectively, one and two orders of magnitude shorter than the time taken by isometric tension to recover 95% of the value before the step after a large release imposed at the tetanus plateau. The overall speed of tension recovery increases continuously from the region of largest releases to that of largest stretches (Fig. 16.8B), in spite of the fact that, for the largest releases, an early faster component of recovery becomes again clearly distinguishable. This last phenomenon is predicted by Huxley and Simmons's theory, since a large step release reduces cross-bridge extension enough to neutralise the effects of superposed lengthening.

The progressive reduction of the extent and speed of phase 2 with increasing lengthening velocity makes it plausible to consider that the recovery during fast lengthening is dominated by the events occurring in phase 4 – that is, cross-bridge detachment and reattachment. Recovery in this phase will be complete when all the cross-bridges attached before the step have detached and reattached further along the actin filament. The finding that the amount of lengthening necessary to recover 95% of the tension before the step is linearly related with the step size (Fig. 16.8C) indicates that the steady state is attained when all the preexisting cross-bridges have been pulled out to a constant degree of strain, slightly above 3 nm per h.s. beyond their original attachment position. Note that 3 nm per h.s. is the amount of cross-bridge strain during slow lengthening evaluated by the shift of the T_2 curve intercept on the length axis. These results show that there is a critical degree of strain (just above 3 nm per h.s.) at which detachment occurs at a very high rate and is followed by a very fast reattachment.

The characteristics of the tension transient during slow lengthening indicate that this forcible detachment occurs at an early stage of the force-generating process. This stage is characterised by a cross-bridge state exhibiting high stiffness as well as high resistance, even to very slow stretches. In biochemical terms, these results imply that, when a cross-bridge is forcibly detached before the release of the products of ATP hydrolysis (presumably before the release of ADP), reattachment is much faster than when it detaches after the release of hydrolysis products, as in isometric conditions or during shortening.

Computer simulation

A computer simulation of the steady-state relation among force, stiffness, and lengthening velocity, as well as of the tension transients elicited either at the isometric tetanus plateau or during steady lengthening at different velocities, was made on the basis of a cross-bridge model which, as in Julian, Sollins, and Sollins (1974), combined the Huxley (1957a) and Huxley and Simmons (1971b) theories. The sequence of cross-bridge reactions assumed in the model is:

$$D1 \underset{k_{-1}}{\overset{k_1}{\rightleftarrows}} A1 \underset{k_{-2}}{\overset{k_2}{\rightleftarrows}} A2 \underset{k_{-3}}{\overset{k_3}{\rightleftarrows}} A3 \xrightarrow{k_4} D1$$

with $A1 \underset{k_{-5}}{\overset{k_5}{\rightleftarrows}} D2 \xrightarrow{k_6}$ (to A2)

The scheme includes two detached (D1 and D2) and three attached (A1, A2, and A3) states. The rate constants for the transitions between attached states and from attached to detached states are given ad hoc functions in relation to cross-bridge extension (Lombardi & Piazzesi, 1990). Under isometric conditions cross-bridge attachment and detachment occur through steps 1 and 4, the rate constants of which are relatively low. The value of the rate constant for detachment becomes high when cross-bridges are negatively strained as a consequence of shortening. The main divergence from the Julian et al. (1974) model consists of a detachment process (steps 5 & 6) that can occur before the completion of the normal cycle and that is made to become relevant when cross-bridges are negatively or positively strained beyond critical values. Reattachment in the case is very fast, being controlled by k_5, which is two orders of magnitude larger than k_1. The force-generating process occurs with change in extension of cross-bridge elastic component during either step 2 or step 3. The increase in extension for each of the two transitions is 5 nm, so that the functions expressing the values of k_3 and k_{-3} are shifted by 5 nm with respect to those for k_2 and k_{-2}. Thus, in isometric conditions, forward reaction 3 is rate limiting, and attached cross-bridges are distributed between A1 and A2.

During steady shortening, a reduction in cross-bridge extension causes a redistribution of cross-bridges towards A3 and, therefore, an increase in rate of detachment. k_1 is moderate so that, with increasing shortening velocity, the number of attached cross-bridges is reduced and the rate of attachment is increased. If we assume that ATP splitting is associated with detachment at the end of the cycle, the increased cross-bridge turnover during shortening will be associated with an increased ATP consumption. The possibility that slightly compressed cross-bridges can detach from state A2 without completion of the normal cycle (step 6) and can reattach at a high rate (through k_5) ensures that both a rise in ATP splitting and a reduction in the number of attached cross-bridges, expected with the increase in shortening speed, are maintained within realistic limits.

During steady lengthening, an increase in cross-bridge extension causes a redistribution of attached cross-bridges towards A1. Beyond a critical amount of cross-bridge strain, a fast detachment to D2, without completion of the normal cycle (and ATP consumption), is provided through steps 5 and 6, whose detachment rate constants rise exponentially with cross-bridge extension. Reattachment through k_5 will also be very fast.

The simulated relations among steady-state tension, stiffness, and lengthening velocity are shown in Fig. 16.9. Stiffness remains practically constant over the whole range of lengthening velocities tested, while the increase in force that accompanies an increase in speed decreases progressively. Certain aspects of the simulation, such as the increase in the number of attachments during lengthening and the conspicuous increase in steady force for lengthening at low speed, are less marked than those found experimentally. It must be noted, in this connexion, that in the model no allowance is made for the possibility that, when filaments

Fig. 16.9. Simulated relations of steady force (T/T_0) and stiffness (S/S_0), expressed relative to the isometric value, versus lengthening velocity. Stiffness was estimated by the total number of attached cross-bridges, assuming that an attached cross-bridge has the same stiffness regardless of its state and extension.

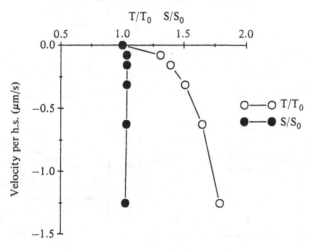

slide, the number of sites available for cross-bridge attachment could increase with respect to the isometric conditions because of the difference in periodicity between sites in the two sets of filaments (H. E. Huxley & Brown, 1967).

The responses of the model for the tension transients following stepwise length changes of different sizes, imposed either at the plateau of the isometric tetanus or during the steady phase of force response to lengthening at different speeds, are shown in Fig. 16.10. It can be seen that the simulated responses give a satisfactory representation of the modifications caused by steady lengthening on the various phases of the transient. The early tension recovery is smaller and slower during slow lengthening (0.15 μm/s per h.s.) than under isometric conditions for small steps (2 & -2 nm per h.s.), but it becomes larger for large releases (-6 & -10 nm per h.s.). Phase 3, where evident (-6 & -10 nm per h.s.), is clearly reduced during slow lengthening. The recovery during

Fig. 16.10. Superposed simulated tension transients elicited by imposing step length changes either at the isometric tetanus plateau (circles) or during steady force response to lengthening at velocities of 0.15 (triangles) and 1.25 (squares) μm/s per h.s. Figures at the top of each graph indicate the size and the sign of the step.

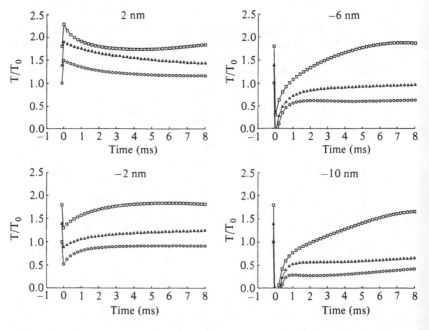

fast lengthening (1.25 μm/s per h.s.) is faster for stepwise stretches than for stepwise releases, but it is, in any case, 95% complete, also for the large releases (-6 & -10 nm per h.s.), within the first 10 ms. At the same time, for large releases an early faster component becomes again clearly distinct from the rest of recovery.

References

Abbott, B. C., & Aubert, X. M (1951). Changes of energy in muscle during very slow stretches. *Proceedings of the Royal Society [B] 139,* 104–17.

Abbott, B. C., & Aubert, X. M. (1952). The force exerted by active striated muscle during and after changes of length. *Journal of Physiology, London, 117,* 77–86.

Abbott, B. C., Aubert, X. M., & Hill, A. V. (1951). The absorption of work by a muscle stretched during a single twitch or a short tetanus. *Proceedings of the Royal Society [B], 139,* 86–104.

Abbott, R. H. (1972). Comments on the mechanism of force generation in striated muscles. *Nature New Biology, 239,* 183–6.

Adelman, Jr., W. J., & Palti, Y. (1969). The effects of external potassium and long duration voltage conditioning on the amplitude of sodium currents in the giant axon of the squid, *Loligo pealei. Journal of General Physiology, 54,* 589–606.

Allen, D. G., & Blinks, J. R. (1978). Calcium transients in aequorin-injected frog cardiac muscle. *Nature, London, 273,* 509–13.

Allen, D. G., & Kurihara, S. (1982). The effect of muscle length on intracellular calcium transients in mammalian cardiac muscle. *Journal of Physiology, London, 327,* 79–94.

Allen, D. G., Lee, J. A., & Westerblad, H. (1989). Intracellular calcium and tension during fatigue in isolated single muscle fibres from *Xenopus laevis. Journal of Physiology, 415,* 433–58.

Allen, R. D. (1986). Video-enhanced microscopy. In *Optical Methods in Cell Physiology,* Vol. 40, ed. P. DeWeer & B. M. Salzberg, pp. 3–13. New York: Wiley.

Allen, T., Barsotti, R. J., Gordon, A. M., Kaplan, J. H., & Goldman, Y. E. (1989). Structural changes in TnC during activation of skinned psoas fibers by laser photolysis of caged Ca. *Biophysical Journal, 55,* 9a.

Allen, T. StC., & Gordon, A. M. (1989). Cycling cross-bridges produce structural changes in troponin-C in demembranated fibres of rabbit psoas muscle. *Journal of Physiology, London, 418,* 63P.

Altringham, J. D., & Bottinelli, R. (1985). The descending limb of the sarcomere length–force relation in single muscle fibres of the frog. *Journal of Muscle Research and Cell Motility, 6,* 585–600.

Altringham, J. D., & Pollack, G. H. (1984). Sarcomere length changes in single frog muscle fibres during tetani at long sarcomere lengths. In *Contractile Mechanisms in Muscle,* ed. G. H. Pollack & H. Sugi, pp. 473–93. New York: Plenum Press.

Ambrogi Lorenzini, C., Colomo, F., & Lombardi, V. (1983). Development of force–velocity relation, stiffness and isometric tension in frog single muscle fibres. *Journal of Muscle Research and Cell Motility, 4,* 177–89.

Amemiya, Y., Iwamoto, H., Kobayashi, T., Sugi, H., Tanaka, H., & Wakabayashi, K. (1988). Time-resolved x-ray diffraction studies on the effect of slow length changes on tetanised frog skeletal muscle. *Journal of Physiology, London, 407,* 231–41.

Anderson, I. L., & Jones, E. W. (1976). Porcine malignant hyperthermia: effect of dantrolene soldium on *in vitro* halothane-induced contraction of susceptible muscle. *Anesthesiology, 44,* 57–61.

Andersson-Cedergren, E. (1959). Ultrastructure of motor end-plate and sarcoplasmic components of mouse skeletal muscle fibre as revealed by three-dimensional reconstruction from serial sections. *Journal of Ultrastructure Research, 10* (Suppl. 1), 1–191.

Araki, M., Takagi, A., Fujita, T., & Matsubara, T. (1985). Porcine malignant hyperthermia: caffeine contracture of single skinned muscle fibers. *Biomedical Research, 6,* 73–8.

Armstrong, C. M., Huxley, A. F., & Julian, F. J. (1966). Oscillatory responses in frog skeletal muscle fibres. *Journal of Physiology, London, 186,* 26–7P.

Astbury, W. T. (1947). On the structure of biological fibres and the problem of muscle. *Proceedings of the Royal Society of London [B], 134,* 303–28.

Aubert, X. (1944a). La chaleur dégagée par un muscle soumis à un travail en cycle. *Comptes Rendus Hebdomadaires des Séances et Mémoires de la Société de Biologie, 138,* 1011–3.

Aubert, X. (1944b). L'Influence du délai de relâchement sur la chaleur dégagée par un muscle soumis à un travail en cycle. *Comptes Rendus Hebdomadaires des Séances et Mémoires de la Société de Biologie, 138,* 1048–50.

Aubert, X. (1948). Réversibilité partielle de la contraction musculaire au cours de l'absorption du travail en cycle. *Archives Internationales de Physiologie, 55,* 348–61.

Austin, K. L., & Denborough, M. A. (1977). Drug treatment of malignant hyperpyrexia. *Anesthesia and Intensive Care, 5,* 207–13.

Bagni, M. A., Cecchi, G., & Schoenberg, M. (1988). A model of force production that explains the lag between crossbridge attachment and force after electrial stimulation of striated muscle fibers. *Biophysical Journal, 54,* 1105–14.

Bagshaw, C. R. (1982). *Muscle Contraction*. London: Chapman & Hall.

Baker, P. F., Blaustein, M. P., Hodgkin, A. L., & Steinhardt, R. A. (1969). The influence of calcium on sodium efflux in squid axons. *Journal of Physiology, London, 200*, 431–58.

Baldwin, T. J., & Burden, S. J. (1989). Muscle-specific gene expression controlled by a regulatory element lacking a MyoD1-binding site. *Nature, London, 341*, 716–20.

Ballard, D. H., & Brown, C. M. (1982). *Computer Vision*. Englewood Cliffs, N.J.: Prentice Hall.

Bárány, K., Bárány, M., Hager, S. R., & Sayers, S. T. (1983). Myosin light chain and membrane protein phosphorylation in various muscles. *Federation Proceedings, 42*, 27–82.

Baylor, S. M., & Hollingworth, S. (1988). Fura-2 calcium transients in frog skeletal muscle fibres. *Journal of Physiology, London, 403*, 151–92.

Baylor, S. M., Chandler, W. K., & Marshall, M. W. (1983). Sarcoplasmic reticulum calcium release in frog skeletal muscle fibres estimated from Arsenazo III calcium transients. *Journal of Physiology, London, 344*, 625–66.

Beam, K. G., Knudson, C. M., & Powell, J. A. (1986). A lethal mutation in mice eliminates the slow calcium current in skeletal muscle cells. *Nature, London, 320*, 168–70.

Bean, B. P., Cohen, C. J., & Tsien, R. W. (1983). Lidocaine block of cardiac sodium channels. *Journal of General Physiology, 81*, 613–42.

Benndorf, K., & Nilius, B. (1987). Inactivation of sodium channels in isolated myocardial mouse cells. *European Biophysical Journal, 15*, 117–27.

Bennett, H. S. (1955). Modern concepts of structure of striated muscle. *American Journal of Physical Medicine, 34*, 46–7.

Bennett, H. S., & Porter, K. R. (1953). An electron microscope study of sectioned breast muscle of the domestic fowl. *American Journal of Anatomy, 93*, 61–105.

Bennett, M. R., Davies, A. M., & Everett, A. W. (1989). The development of topographical maps and fibre types in toad (*Bufo marinus*) gluteus muscle during synapse elimination. *Journal of Physiology, London, 409*, 43–61.

Bergman, R. A. (1983). Ultrastructural configuration of sarcomeres in passive and contracted frog sartorius muscle. *American Journal of Anatomy, 166*, 209–22.

Bernstein, J. (1902). Untersuchungen zur Thermodynamik der bioelektrischen Ströme. Erster Theil. *Pflügers Archiv, 92*, 521–62.

Bershitsky, S. U., & Tsaturyan, A. K. (1989). Effect of joule temperature jump on tension and stiffness of skinned rabbit muscle fibers. *Biophysical Journal, 56*, 809–16.

Bezanilla, F., & Armstrong, C. M. (1977). Inactivation of the sodium channel. I. Sodium current experiments. *Journal of General Physiology, 70*, 549–66.

Bezanilla, F., Caputo, C., Gonzalez-Serratos, H., & Venosa, R. A. (1972).

260   *References*

Sodium dependence of the inward spread of activation in isolated twitch muscle fibres of the frog. *Journal of Physiology, London, 223,* 509–23.

Bianchi, C. P., & Narayan, S. (1982). Muscle fatigue and the role of transverse tubules. *Science, 215,* 295–6.

Bigland-Ritchie, B., & Woods, J. J. (1984). Changes in muscle contractile properties and neural control during human muscular fatigue. *Muscle and Nerve, 7,* 691–9.

Blangé, T., Karemaker, J. M., & Kramer, A. E. J. L. (1972). Tension transients after quick release in rat and frog skeletal muscle. *Nature, London, 237,* 281–2.

Blaschko, H., Cattell, M., & Kahn, J. L. (1931). On the nature of the two types of response in the neuromuscular system of the crustacean claw. *Journal of Physiology, London, 73,* 25–35.

Blinks, J. R. (1965). Influence of osmotic strength on cross-section and volume of isolated single muscle fibres. *Journal of Physiology, London, 177,* 42–57.

Blinks, J. R., & Endoh M. (1986). Modification of myofibrillar responsiveness to Ca^{++} as an inotropic mechanism. *Circulation, 73* (Suppl. III). III85–98.

Blinks, J. R., Rüdel, R., & Taylor, S. R. (1978). Calcium transients in isolated amphibian skeletal muscle fibres: detection with aequorin. *Journal of Physiology, London, 277,* 291–323.

Bliton, A. C., Patton, M. J., Rolli, M. L., Roos, K. P., & Taylor, S. R. (1988). Microscopic motion analysis, Laplacian–Gaussian masks for subpixel edge detection. *Proceedings of the Annual International Conference of the IEEE Engineering in Medicine and Biology Society, 10,* 1098–9.

Block, B. A., Imagawa, T., Campbell, K. P., & Franzini-Armstrong, C. (1988). Structural evidence for direct interaction between the molecular components of the transverse/tubule sarcoplasmic reticulum junction in skeletal muscle. *Journal of Cell Biology, 107,* 2587–600.

Boehm, R. (1914). Über das Verhalten des isolierten Froschherzens bei reiner Salzdiät. *Archiv für Experimentelle Pathologie und Pharmakologie, 75,* 230–316.

Boldin, S., Jager, U., Ruppersberg, J. P., Pentz, S., & Rüdel, R. (1987). Cultivation, morphology, and electrophysiology of contractile rat myoballs. *Pflügers Archiv, 409,* 462–7.

Bowditch, H. P. (1871). Über die Eigenthümlichkeiten der Reizbarkeit, welche die Muskelfasern des Herzens zeigen. *Berichte Sächsische Gesellschaft (Akademie) der Wissenschaften, 23,* 652–89.

Bradbury, S. (1989). Landmarks in biological light microscopy. *Journal of Microscopy, 155,* 281–305.

Brandl, C. J., Green, N. M., Korczak, B., & MacLennan, D. H. (1986). Two Ca^{2+} ATPase genes: homologies and mechanistic implications of deduced amino acid sequences. *Cell, 44,* 597–607.

Brandt, P. W., Diamond, M. S., Rutchik, J. S., & Schachat, F. H. (1987).

Cooperative interactions between troponin–tropomyosin units extend the length of the thin filament in skeletal muscle. *Journal of Molecular Biology*, *195*, 885–96.

Bremel, R. D., & Weber, A. (1972). Cooperation within actin filaments in vertebrate skeletal muscle. *Nature New Biology*, *239*, 97–101.

Brenner, B. (1980). Effect of free sarcoplasmic Ca^{2+} concentration on maximum unloaded shortening velocity: measurements on single glycerinated rabbit psoas muscle fibres. *Journal of Muscle Research and Cell Motility*, *1*, 409–28.

Brenner, B. (1983). Technique for stabilising the striation pattern in maximally calcium-activated skinned rabbit psoas fibers. *Biophysical Journal*, *41*, 99–102.

Brenner, B. (1986). The necessity of using two parameters to describe isotonic shortening velocity of muscle tissues; the effect of various interventions upon initial shortening velocity (V_i) and curvature (b). *Basic Research in Cardiology*, *6*, 54–89.

Brenner, B. (1988a). Effect of Ca^{2+} on cross-bridge turnover kinetics in skinned single rabbit psoas fibers: implications for regulation of muscle contraction. *Proceedings of the National Academy of Sciences, USA*, *85*, 3265–9.

Brenner, B. (1988b). Evidence for rapidly reversible cross-bridge attachment to actin in Ca^{++}-activated skinned rabbit psoas fibers. *Pflügers Archiv*, *412* (Suppl. 1), R79.

Brenner, B. (1990). Transient detachment of force-generating cross-bridges in Ca^{2+}-activated skinned psoas fibres of the rabbit. *Journal of Physiology, London*, *426*, 40P.

Brenner, B., Maeda, Y., & Simmons, R. M. (1990). Effect of isotonic shortening on the equatorial x-ray diffraction pattern of skinned single rabbit psoas fibers. *Biophysical Journal*, *57*, 539a.

Brenner, B., Schoenberg, M., Chalovich, J. M., Greene, L. E., & Eisenberg, E. (1982). Evidence for cross-bridge attachment in relaxed muscle at low ionic strength. *Proceedings of the National Academy of Sciences, USA*, *79*, 7288–91.

Brismar, T. (1977). Slow mechanism of sodium permeability inactivation in myelinated nerve fibre of *Xenopus laevis*. *Journal of Physiology, London*, *270*, 283–97.

Brocklehurst, L. (1975). Dantrolene sodium and "skinned" muscle fibres. *Nature, London*, *254*, 364.

Brown, L. M. (1975a). A biologist's guide to the theory of optical diffraction. In *Structural Changes in Striated Muscle Fibres Shortened Passively Below Their Slack Length*. Ph.D. thesis, University of London, Appendix I, pp. 254–88.

Brown, L. M. (1975b). Calibration of the Siemens Elmiskop I. In *Structural Changes in Striated Muscle Fibres Shortened Passively Below Their Slack Length*. Ph.D. thesis, University of London, Appendix II, pp. 289–307.

Brown, L. M. (1978). Calibration of a commercial electron microscope with a grating replica to an accuracy of better than 1%. *Journal of Microscopy, 113,* 149–60.

Brown, L. M., & Farquharson, D. B. (1984). A simple method for making accurate measurements from electron micrograph negatives. *Journal of Physiology, London, 353,* 13P.

Brown, L. M., Gonzalez-Serratos, H., & Huxley, A. F. (1970). Electron microscopy of frog muscle fibres in extreme passive shortening. *Journal of Physiology, London, 208,* 86–8P.

Brown, L. M., Gonzalez-Serratos, H., & Huxley, A. F. (1984a). Structural studies of the waves in striated muscle fibres shortened passively below their slack length. *Journal of Muscle Research and Cell Motility, 5,* 273–92.

Brown, L. M., Gonzalez-Serratos, H., & Huxley, A. F. (1984b). Sarcomere and filament lengths in passive muscle fibres with wavy myofibrils. *Journal of Muscle Research and Cell Motility, 5,* 293–314.

Brown, L. M., & Hill, L. (1982). Mercuric chloride in alcohol and chloroform used as a rapidly acting fixative for contracting muscle fibres. *Journal of Microscopy, 125,* 319–36.

Brown, L. M., & Hill, L. (1991). Some observations on variations in filament overlap in tetanised muscle fibres and fibres stretched during a tetanus, detected in the electron microscope after rapid fixation. *Journal of Muscle Research and Cell Motility, 12,* 171–82.

Brown, L. M., Lopez, J. R., Olsen, J. A., Rüdel, R., Simmons, R. M., Taylor, S. R., & Wanek, L. A. (1982). Branched skeletal muscle fibres not associated with dysfunction. *Muscle and Nerve, 5,* 645–53.

Brutsaert, D. L., Claes, V. A., & Sonnenblick, E. H. (1971). Effects of abrupt load alterations on force–velocity–length and time relations during isotonic contractions of heart muscle: load clamping. *Journal of Physiology, London, 216,* 319–330.

Burton, K. (1989). The magnitude of the force rise after rapid shortening and restretch in fibres isolated from rabbit psoas muscle. *Journal of Physiology, London, 418,* 66P.

Burton, K., & Simmons, R. M. (1990). The transition from the isometric state to steady shortening in rabbit skinned fibers. *Biophysical Journal, 57,* 548a.

Campbell, D. L., Giles, W. R., Robinson, K., & Shibata, E. F. (1988a). Studies of the sodium-calcium exchanger in bull-frog atrial myocytes. *Journal of Physiology, London, 403,* 317–40.

Campbell, D. L., Giles, W. R., & Shibata, E. F. (1988b). Ion transfer characteristics of the calcium current in bull-frog atrial myocytes. *Journal of Physiology, London, 403,* 239–66.

Campbell, K. P., Franzini-Armstrong, C., & Shamoo, A. (1980). Further characterization of light and heavy sarcoplasmic reticulum vesicles. Identification of the 'sarcoplasmic reticulum feet' associated with heavy sarcoplasmic reticulum vesicles. *Biochimica et Biophysica Acta, 602,* 97–116.

Carlsen, F., Knappeis, G. G., & Buchtal, F. (1961). Ultrastructure of the resting and contracted striated muscle fiber at different degrees of stretch. *Journal of Biophysical and Biochemical Cytology, 11*, 95–117.

Caspersson, T., & Thorell, B. (1942). The localization of the adenylic acids in striated muscle-fibres. *Acta Physiologica Scandinavica, 4*, 97–117.

Caswell, A. H., & Brandt, N. R. (1989). Does muscle activation occur by direct mechanical coupling of transverse tubules of sarcoplasmic reticulum? *Trends in Biochemical Science, 14*, 161–5.

Caswell, A. H., & Brunschwig, J. P. (1984). Identification and extraction of proteins that compose the triad junction of skeletal muscle. *Journal of Cell Biology, 99*, 929–39.

Cecchi, G., Colomo, F., & Lombardi, V. (1978). Force–velocity relation in normal and nitrate-treated frog single muscle fibres during rise of tension in an isometric tetanus. *Journal of Physiology, London, 5*, 257–73.

Cecchi, G., Colomo, F., Lombardi, V., & Piazzesi, G. (1987). Stiffness of frog muscle fibres during rise of tension and relaxation in fixed-end or length-clamped tetani. *Plügers Archiv, 409*, 39–46.

Cecchi, G., Griffiths, P. J., & Taylor, S. (1982). Muscular contraction: kinetics of crossbridge attachment studied by high-frequency stiffness measurements. *Science, 217*, 70–2.

Cecchi, G., Griffiths, P. J., & Taylor, S. (1986). Stiffness and force in activated frog skeletal muscle fibers. *Biophysical Journal. 49*, 437–51.

Chapman, R. A., Coray, A., & McGuigan, J. A. S. (1983). Sodium/calcium exchange in mammalian ventricular muscle: a study with sodium sensitive micro-electrodes. *Journal of Physiology, London, 343*, 253–76.

Chapman, R. A., & Niedergerke, R. (1970a). Effects of calcium on the contraction of the hypodynamic frog heart. *Journal of Physiology, London, 211*, 389–421.

Chapman, R. A., & Niedergerke, R. (1970b). Interaction between heart rate and calcium concentration in the control of contractile strength of the frog heart. *Journal of Physiology, London, 211*, 423–43.

Chapman, R. A., & Tunstall, J. (1984). The measurement of intracellular sodium activity and its relationship to the atrial trabeculae. *Quarterly Journal of Experimental Physiology, 69*, 559–72.

Cheung, J. Y., Tillotson, D. L., Yelmarty, R. V., & Scaduto, Jr., R. C. (1989). Cytosolic free calcium concentration in individual cardiac myocytes in primary culture. *American Journal of Physiology, 256*, C1120–30.

Chiu, S. Y. (1977). Inactivation of sodium channels: second order kinetics in myelinated nerve. *Journal of Physiology, London, 273*, 573–96.

Chiu, Y. C., Ballou, E. W., & Ford, L. E. (1987). Force, velocity, and power changes during normal and potentiated contractions of cat papillary muscle. *Circulation Research, 60*, 446–58.

Chiu, Y. C., Walley, K. R., & Ford, L. E. (1989). Comparison of the effects of different inotropic interventions on force velocity and power in rabbit myocardium. *Circulation Research, 65,* 1161–71.

Chiu, Y-L., Ballou, E. W., & Ford, L. E. (1982a). Internal viscoelastic loading in cat papillary muscle. *Biophysical Journal, 40,* 109–20.

Chiu, Y-L., Ballou, E. W., & Ford, L. E. (1982b). Velocity transients and viscoelastic resistance to active shortening in cat papillary muscle. *Biophysical Journal, 40,* 121–8.

Civan, M. M., & Podolsky, R. J. (1966). Contraction kinetics of striated muscle fibres following quick changes in load. *Journal of Physiology, London, 184,* 511–34.

Clark, A. J. (1913). The action of ions and lipoids upon the frog's heart. *Journal of Physiology, London, 47,* 66–107.

Cole, H., Frearson, N., Moir, A., Perry, S. V., & Solaro, J. (1978). Phosphorylation of cardiac myofibrillar proteins in heart function and metabolism. *Recent Advances in Studies on Cardiac Structure and Metabolism, 11,* 111–20.

Colomo, F., Lombardi, V., Menchetti, G., & Piazzesi, G. (1989a). The recovery of isometric tension after steady lengthening in tetanised fibres isolated from frog muscle. *Journal of Physiology, London, 415,* 130P.

Colomo, F., Lombardi, V., & Piazzesi, G. (1988). The mechanisms of force enhancement during constant velocity lengthening in tetanised single fibres of frog muscle. In *Molecular Mechanisms of Muscle Contraction,* ed. H. Sugi & G. H. Pollack, pp. 489–502. New York: Plenum Press.

Colomo, F., Lombardi, V., & Piazzesi, G. (1989b). The recovery of tension in transients during steady lengthening of frog muscle fibres. *Pflügers Archiv, 414,* 245–7.

Cooke, R., Crowder, M. S., Wendt, C. H., Barnett, V. A., & Thomas, D. D. (1984). Muscle cross-bridges: do they rotate? In *Contractile Mechanisms in Muscle,* ed. G. H. Pollack & H. Sugi, pp. 413–27. New York and London: Plenum Press.

Cooke, R., Franks, K., Luciani, G., & Pate, E. (1988). The inhibition of rabbit skeletal muscle contraction by hydrogen ions and phosphate. *Journal of Physiology, London, 395,* 77–97.

Corbett, A. M., Caswell, A. H., Brandt, N. R., & Brunschwig, J. P. (1985). Determinants of triad junction reformation: identification and isolation of an endogenous promotor for junction reformation in skeletal muscle. *Journal of Membrane Biology, 86,* 267–76.

Constantin, L. L. (1970). The role of sodium current in the radial spread of contraction in frog muscle fibers. *Journal of General Physiology, 50,* 703–15.

Costantin, L. L., & Taylor, S. R. (1973). Graded activation in frog muscle fibers. *Journal of General Physiology, 61,* 424–43.

Craig, R., & Offer, G. (1976). Axial arrangement of crossbridges in thick filaments of vertebrate striated muscle. *Journal of Molecular Biology, 102,* 325–32.

Curtin, N. A., & Davies, R. E. (1973). Chemical and mechanical changes during stretching of activated frog skeletal muscle. *Cold Spring Harbor Symposia on Quantitative Biology, 37,* 619–26.

Curtin, N. A., Howarth, J. V., & Woledge, R. C. (1983). Heat production by single fibres from frog muscle. *Journal of Muscle Research and Cell Motility, 4,* 207–22.

Curtin, H. J., & Cole, K. S. (1940). Membrane action potentials from the squid giant axon. *Journal of Cellular and Comparative Physiology, 15,* 147–57.

Curtis, H. J., & Cole, K. S. (1942). Membrane resting and action potentials from the squid giant axon. *Journal of Cellular and Comparative Physiology, 19,* 135–44.

Dantzig, J. A., & Goldman, Y. E. (1989). Tension and stiffness decline following photolysis of caged phosphate within actively contracting glycerol-extracted fibers of rabbit psoas muscle. *Biophysical Journal, 55,* 11a.

Dantzig, J. A., Hibberd, M. G., Trentham, D. R., & Goldman, Y. E. (1991). Cross-bridge kinetics in the presence of MgADP investigated by photolysis of caged ATP in rabbit psoas muscle fibres. *Journal of Physiology, London, 432,* 639–80.

Dantzig, J. A., Lacktis, J. W., Homsher, E., & Goldman, Y. E. (1987). Mechanical transients initiated by photolysis of caged P_i during active skeletal muscle contractions. *Biophysical Journal, 51,* 3a.

Davis, J. S., & Harrington, W. F. (1987). Force generation by muscle fibers in rigor: a laser temperature-jump study. *Proceedings of the National Academy of Sciences, USA, 84,* 975–9.

Dawson, M. J., Gadian, D. G., & Wilkie, D. R. (1978). Muscular fatigue investigated by phosphorus nuclear magnetic resonance. *Nature, London, 274,* 861–6.

Dawson, M. J., Gadian, D. G., & Wilkie, D. R. (1980). Mechanical relaxation rate and metabolism studied in fatiguing muscle by phosphorus nuclear magnetic resonance. *Journal of Physiology, London, 299,* 465–84.

Deitmer, J. W., & Ellis, D. (1978). Changes in the intracellular sodium activity of sheep heart Purkinje fibres produced by calcium and other divalent cations. *Journal of Physiology, London, 277,* 437–53.

Delay, M. J., Ishide, N., Jacobson, R. C., Pollack, G. H., & Tirosh, R. (1981). Stepwise sarcomere shortening: analysis by high-speed cinemicrography. *Science, 213,* 1523–5.

Denborough, M. A., & Lovell, R. R. H. (1960). Anesthetic deaths in a family. *Lancet, II,* 45.

Desmedt, J. E., & Hainaut, K. (1977). Inhibition of the intracellular release of calcium by dantrolene in barnacle giant muscle fibres. *Journal of Physiology, London, 265,* 565–85.

DiFrancesco, D., & Noble, D. (1985). A model of cardiac electrical activity incorporating ionic pumps and concentration changes. *Philosophical Transactions of the Royal Society, London [B] 307*, 353–98.

DiPolo, R. (1979). Calcium influx in internally dialyzed squid giant axons. *Journal of General Physiology, 73*, 91–113.

DiPolo, R., & Beaugé, L. (1987). Characterization of the reverse Na/Ca exchange in squid axons and its modulation by Ca_i and ATP. *Journal of General Physiology, 90*, 505–25.

Dudel, J., Peper, K., Rüdel, R., & Trautwein, W. (1966). Excitatory membrane current in heart muscle (Purkinje fibres). *Pflügers Archiv, 292*, 255–73.

Dudel, J., & Rüdel, R. (1970). Voltage and time dependence of excitatory sodium current in cooled sheep Purkinje fibres. *Pflügers Archiv, 315*, 136–58.

Dulhunty, A. F. (1989). Feet, bridges and pillars in triad junctions of mammalian skeletal muscle, their possible relationship to calcium buffers in terminal cisternae and T-tubules and to excitation–contraction coupling. *Journal of Membrane Biology, 109*, 73–83.

Ebashi, S., Ebashi, F., & Kodama, A. (1967). Troponin as the Ca^{++}-receptive protein in the contractile system. *Journal of Biochemistry, Tokyo, 62*, 137–8.

Ebashi, S., & Endo, M. (1968). Calcium ion and muscle contraction. *Progress in Biophysics & Molecular Biology*, 123–83.

Eberstein, A., & Sandow, A. (1963). Fatigue mechanisms in muscle fibres. In *The Effect of Use and Disuse on Neuromuscular Functions*, ed. E. Gutman & P. Haik, pp. 515–26. Amsterdam: Elsevier.

Edman, K. A. P. (1975). Mechanical deactivation induced by active shortening in isolated muscle fibres of the frog. *Journal of Physiology, London, 246*, 255–75.

Edman, K. A. P., Elzinga, G., & Noble, M. I. M. (1978). Enhancement of mechanical performance by stretch during tetanic contractions of vertebrate skeletal muscle fibres. *Journal of Physiology, London, 281*, 139–55.

Edman, K. A. P., Elzinga, G., & Noble, M. I. M. (1982). Residual force enhancement after stretch of contracting frog single muscle fibres. *Journal of General Physiology, 80*, 769–84.

Edman, K. A. P., Elzinga, G., & Noble, M. I. M. (1984). Stretch of contracting muscle fibres: evidence for regularly spaced active sites along the filaments and enhanced mechanical performance. In *Contractile Mechanisms in Muscle*, ed. G. H. Pollack & H. Sugi, pp. 739–51. New York: Plenum Press.

Edman, K. A. P., & Mattiazi, A. R. (1981). Effects of fatigue and altered pH on isometric force and velocity of shortening at zero load in frog muscle fibres. *Journal of Muscle Research and Cell Contractility, 2*, 321–34.

Edman, K. A. P., & Reggiani, C. (1984). Redistribution of sarcomere length during isometric contraction of frog muscle fibres and its relation to tension creep. *Journal of Physiology, London, 351*, 169–98.

Edwards, R. H. T. (1981). Human muscle function and fatigue. In *Human*

Muscle Fatigue: Physiological Mechanisms. Ciba Found. Symp., *82*, ed. R. Porter & J. Whelan, pp. 1–18. London: Pitman Medical.

Edwards, R. H. T., Hill, D. K., & Jones, D. A. (1975). Metabolic changes associated with the slowing of relaxation in fatigued mouse muscle. *Journal of Physiology, London*, *251*, 303–15.

EG&G Reticon (1989). *Solid State Camera Products*. Sunnyvale, EG&G Reticon.

Eisenberg, B. R., & Eisenberg, R. S. (1982). The T–SR junction in contracting single skeletal muscle fibres. *Journal of General Physiology*, *79*, 1–19.

Eisenberg, E., & Greene, L. (1980). The relation of muscle biochemistry to muscle physiology. *Annual Review of Physiology*, *42*, 293–309.

Eisenberg, E., & Hill, T. L. (1985). Muscle contraction and free energy transduction in biological systems. *Science*, *227*, 999–1006.

Eisner, D. A., Lederer, W. J., & Vaughan-Jones, E. D. (1983). The control of tonic tension by membrane potential and intracellular sodium activity in the sheep cardiac Purkinje fibre. *Journal of Physiology, London*, *335*, 723–43.

Eisner, D. A., Lederer, W. J., & Vaughan-Jones, R. D. (1984). The quantitative relationship between twitch tension and intracellular sodium activity in sheep cardiac Purkinje fibres. *Journal of Physiology, London*, *355*, 251–66.

Elliott, J. R., Haydon, D. A., & Henry, B. M. (1987). The mechanisms of sodium current inhibition by benzocaine in the squid giant axon. *Pflügers Archiv*, *409*, 596–600.

Ellis, D. (1977). The effects of external cations and ouabain on the intracellular sodium activity of sheep heart Purkinje fibres. *Journal of Physiology, London*, *273*, 211–40.

Ellis, K. O., & Bryant, S. H. (1972). Excitation contraction uncoupling in skeletal muscle by dantrolene sodium. *Naunyn Schmiedeberg's Archives of Pharmacology*, *274*, 107–9.

Ellis, K. O., & Carpenter, J. F. (1972). Studies on the mechanism of action of dantrolene sodium. *Naunyn Schmiedeberg's Archives of Pharmacology*, *275*, 83–94.

El-Saleh, S. C., Warber, K. D., & Potter, J. D. (1986). The role of tropomyosin-troponin in the regulation of skeletal muscle contraction. *Journal of Muscle Research and Cell Motility*, *7*, 387–404.

Elzinga, G., Lännergren, J., & Stienen, G. J. M. (1987). Stable maintenance heat rate and contractile properties of different single muscle fibres from *Xenopus laevis* at 20°C. *Journal of Physiology, London*, *393*, 399–412.

Elzinga, G., & van der Laarse, W. J. (1988). Oxygen consumption of single muscle fibres of *Rana temporaria* and *Xenopus laevis* at 20°C. *Journal of Physiology, London*, *399*, 405–18.

Endo, M. (1964). Entry of a dye into the sarcotubular system of muscle. *Nature, London*, *202*, 1115–16.

Endo, M. (1966). Entry of fluorescent dyes into the sarcotubular system of the frog muscle. *Journal of Physiology, London, 185,* 224–38.

Endo, M. (1977). Calcium release from the sarcoplasmic reticulum. *Physiological Reviews, 57,* 71–108.

Endo, M. (1981). Mechanism of calcium-induced calcium release in the SR membrane. In *The Mechanism of Gated Calcium Transport Across Biological Membranes,* ed. S. T. Ohnishi & M. Endo, pp. 257–64. New York: Academic Press.

Endo, M. (1984). Significance of calcium-induced release of calcium from the sarcoplasmic reticulum in skeletal muscle. *Jikeikai Medical Journal, 30,* 123–30.

Endo, M. (1985). Calcium release from sarcoplasmic reticulum. *Current Topics in Membranes and Transport, 25,* 181–230.

Endo, M., & Iino, M. (1988). Measurement of Ca^{2+} release in skinned fibers from skeletal muscle. *Methods in Enzymology, 157,* 12–26.

Endo, M., Kakuta, Y., & Kitazawa, T. (1981). A further study of the Ca-induced Ca release mechanism. In *The Regulation of Muscle Contraction: Excitation–Contraction Coupling,* ed. A. D. Grinell & M. A. Brazier, pp. 181–94. New York: Academic Press.

Endo, M., Tanaka, M., & Ebashi, S. (1968). Release of calcium from sarcoplasmic reticulum in skinned fibers of the frog. *Proceedings of the International Union of Physiological Sciences, XXIV International Congress, 7,* 126.

Endo, M., Tanaka, M., & Ogawa, Y. (1970). Calcium induced release of calcium from the sarcoplasmic reticulum of skinned skeletal muscle fibres. *Nature, London, 228,* 34–6.

Endo, M., Yagi, S., Ishizuka, T., Horiuti, K., Koga, Y., & Amaha, K. (1983). Changes in the Ca-induced Ca release mechanism in the sarcoplasmic reticulum of muscle from a patient with malignant hyperthermia. *Biomedical Research, 4,* 83–92.

England, P. (1976). Studies on the phosphorylation of the inhibitory subunit of troponin during modification of contraction in perfused rat heart. *Journal of Biochemistry, 160,* 295–304.

Fabiato, A. (1985). Time and calcium dependence of activation and inactivation of calcium-induced release of calcium from the sarcoplasmic reticulum of a skinned canine cardiac Purkinje cell. *Journal of General Physiology, 85,* 247–89.

Fabiato, A., & Fabiato, F. (1978). Effects of pH on the myofilament and the sarcoplasmic reticulum of skinned cells from cardiac and skeletal muscle. *Journal of Physiology, London, 276,* 233–55.

Fakler, B., Ruppersberg, J. P., Spittelmeister, W., & Rüdel, R. (1990). Inactivation of human sodium channels and the effect of tocainide. *Pflügers Archiv, 415,* 693–700.

Farrow, A. J., Rossmanith, G. H., Unsworth, J. (1988). The role of calcium ions in the activation of rabbit psoas muscle. *Journal of Muscle Research and Cell Motility, 9*, 261–74.

Fenn, W. O. (1923). A quantitative comparison between the energy liberated and the work performed by the isolated sartorius muscle of the frog. *Journal of Physiology, London, 58*, 175–203.

Fenn, W. O. (1924). The relation between the work performed and the energy liberated in muscular contraction. *Journal of Physiology, London, 58*, 373–95.

Ferenczi, M. A. (1986). Phosphate burst in permeable muscle fibers of the rabbit. *Biophysical Journal, 50*, 471–7.

Ferenczi, M. A., Goldman, Y. E., & Simmons, R. M. (1984). The dependence of force and shortening velocity on substrate concentration in skinned muscle fibres from *Rana temporaria. Journal of Physiology, London, 350*, 519–43.

Ferenczi, M. A., Homsher, E., & Trentham, D. R. (1984). The kinetics of magnesium adenosine triphosphate cleavage in skinned muscle fibres of the rabbit. *Journal of Physiology, London, 352*, 575–99.

Ferguson, D. G., & Franzini-Armstrong, C. (1988). The Ca^{2+} ATPase content of slow and fast twitch fibers of guinea pig. *Muscle and Nerve, 11*, 561–70.

Ferguson, D. G., Schwartz, H., & Franzini-Armstrong, C. (1984). Subunit structure of junctional feet in triads of skeletal muscle: a freeze-drying, rotary-shadowing study. *Journal of Cell Biology, 99*, 1735–42.

Fischman, D. A. (1986). Myofibrillogenesis and the morphogenesis of skeletal muscle. In *Myology: Basic and Clinical*, ed. A. G. Engel & B. Q. Banker, pp. 5–37. New York: McGraw-Hill.

Fitts, R., & Holloszy, J. (1976). Lactate and contractile force in frog muscle during development of fatigue and recovery. *American Journal of Physiology, 231*, 430–3.

Fleischer, S., Ogunbunmi, E. M., Dixon, M. C., & Fleer, E. A. (1985). Localization of Ca^{2+} release channels with ryanodine in junctional terminal cisternae of sarcoplasmic reticulum of fast skeletal muscle. *Proceedings of the National Academy of Sciences, USA, 82*, 7256–9.

Flitney, F. W., & Hirst, D. G. (1978). Crossbridge detachment and sarcomere "give" during stretch of active frog's muscle. *Journal of Physiology, London, 276*, 449–65.

Ford, L. E., & Huxley, A. F. (1973). A moving coil device giving step displacements of 0.2 mm in 0.3 ms. *Journal of Physiology, London, 232*, 22P (title only).

Ford, L. E., Huxley, A. F., & Simmons, R. M. (1977). Tension responses to sudden length change in stimulated frog muscle fibres near slack length. *Journal of Physiology, London, 269*, 441–515.

Ford, L. E., Huxley, A. F., & Simmons, R. M. (1981). The relation between stiffness and filament overlap in stimulated frog muscle fibres. *Journal of Physiology, London, 311*, 219–49.

Ford, L. E., Huxley, A. F., & Simmons, R. M. (1985). Tension transients during steady shortening of frog muscle fibres. *Journal of Physiology, London, 361,* 131–50.

Ford, L. E., Huxley, A. F., & Simmons, R. M. (1986). Tension transients during the rise of tetanic tension in frog muscle fibres. *Journal of Physiology, London, 372,* 595–609.

Ford, L., & Podolsky, R. J. (1968). Activation mechanisms of skinned muscle fibers. *Proceedings of the International Union of Physiological Sciences, XXIV International Congress, 7,* 139.

Ford, L., & Podolsky, R. J. (1970). Regenerative calcium release within muscle cells. *Science, 167,* 58–9.

Ford, L., & Podolsky, R. J. (1972a). Calcium uptake and force development by skinned muscle fibres in EGTA buffered solutions. *Journal of Physiology, London, 223,* 1–19.

Ford, L., & Podolsky, R. J. (1972b). Interacellular calcium movements in skinned muscle fibres. *Journal of Physiology, London, 223,* 21–33.

Fox, J. M. (1976). Ultra-slow inactivation of the ionic currents through the membrane of myelinated nerve. *Biochimica et Biophysica Acta, 426,* 232–44.

Frampton, J. E., Orchard, C. H., & Boyett, M. R. (1990). The relationship between diastolic and systolic intracellular $[Ca^{2+}]([Ca^{2+}]_i)$ in ventricular myocytes isolated from rat hearts. *Journal of Physiology, London, 425,* 53P.

Frank, G. (1928). Das histologische Bild der Muskelkontraktion. *Pflügers Archiv, 218,* 37–53.

Frankenhäuser, B., & Huxley, A. F. (1964). The action potential in the myelinated nerve fibre of *Xenopus laevis* as computed on the basis of voltage clamp date. *Journal of Physiology, London, 171,* 302–15.

Franzini-Armstrong, C. (1966). Variations in the length of crab myofilaments. *Journal of Physiology, London, 186,* 25P.

Franzini-Armstrong, C. (1970a). Studies on the triad. I. Structure of the junction of frog twitch fibers. *Journal of Cell Biology, 47,* 488–99.

Franzini-Armstrong, C. (1970b). Natural variability in the length of thin and thick filaments in single fibres from a crab, *Portunus depurator. Journal of Physiology, London, 186,* 25P.

Franzini-Armstrong, C. (1984). Freeze-fracture of frog slow tonic fibers. Structures of surface and internal membranes. *Tissue and Cell, 16,* 647–64.

Franzini-Armstrong, C., & Ferguson, D. G. (1985). Density and disposition of Ca^{2+}-ATPase in sarcoplasmic reticulum membrane as determined by shadowing techniques. *Biophysical Journal, 48,* 607–15.

Franzini-Armstrong, C., Kenney, L. J., & Varriano-Marston, E. (1987). The structure of calsequestrin in triads of vertebrate skeletal muscle: a deep-etch study. *Journal of Cell Biology, 105,* 49–56.

Franzini-Armstrong, C., & Porter, K. R. (1964). Sarcolemmal invaginations constituting the T system in fish muscle fibers. *Journal of Cell Biology, 22,* 675–96.

Frearson, N., Focant, B. W. W., & Perry, S. V. (1976). Phosphorylation of a light chain component of myosin from smooth muscle. *FEBS Letters, 63,* 27–32.

Fridén, J. (1984). Changes in human skeletal muscle induced by long-term eccentric exercise. *Cell and Tissue Research, 236,* 365–72.

Fuchs, F. (1977). The binding of calcium to glycerinated muscle fibers in rigor: the effect of filament overlap. *Biochimica et Biophysica Acta, 491,* 523–31.

Fuchs, F., Reddy, Y., & Briggs, F. N. (1970). The interaction of cations with calcium-binding site of troponin. *Biochimica et Biophysica Acta, 221,* 407–9.

Fukuda, J., Henkart, M. P., Fischbach, G. D., & Smith, T. G. (1976). Physiological and structural properties of colchicine-treated chick skeletal muscle cells grown in tissue culture. *Developmental Biology, 49,* 395–411.

Gadsby, D. C., Niedergerke, R., & Page, S. (1971). Do intracellular concentrations of potassium or sodium regulate the strength of the heart beat? *Nature, London, 232,* 651–3.

Garcia, M. del C., & Gonzalez-Serratos, H. (1982). Caffeine contractions in fatigued twitch muscle fibres. *Biophysical Journal, 37,* 21a.

Garvey, L., Kranias, E., & Solaro, J. (1988). Phosphorylation of C-protein, troponin I and phospholamban in isolated rabbit hearts. *Biochemical Journal, 249,* 709–14.

Gelfan, S. (1933). The submaximal responses of a single muscle fibre. *Journal of Physiology, London, 80,* 285–95.

Glitsch, H. G., Reuter, H., & Scholz, H. (1970). The effect of the internal sodium concentration on calcium fluxes in isolated guinea-pig auricles. *Journal of Physiology, London, 209,* 25–43.

Godt, R. E., & Maughan, D. W. (1977). Swelling of skinned muscle fibers of the frog: experimental observations. *Biophysical Journal, 19,* 103–16.

Godt, R. E., & Nosek, T. M. (1989). Changes of intracellular milieu with fatigue or hypoxia depress contraction of skinned rabbit skeletal and cardiac muscle. *Journal of Physiology, London, 412,* 155–80.

Goldman, L., & Kenyon, J. L. (1982). Delays in inactivation development and activation kinetics in *Myxicola* giant axons. *Journal of General Physiology, 80,* 83–102.

Goldman, Y. E. (1987a). Kinetics of the actomyosin ATPase in muscle fibers. *Annual Review of Physiology, 49,* 637–54.

Goldman, Y. E. (1987b). Measurement of sarcomere shortening in skinned fibers from frog muscle by white light diffraction. *Biophysical Journal, 52,* 57–68.

Goldman, Y. E., Hibberd, M. G., McCray, J. A., & Trentham, D. R. (1982). Relaxation of muscle fibres by photolysis of caged ATP. *Nature, London, 300,* 701–5.

Goldman, Y. E., Hibberd, M. G., & Trentham, D. R. (1984a). Relaxation of rabbit psoas muscle fibres from rigor by photochemical generation of adenosine-5'-triphosphate. *Journal of Physiology, London, 354,* 577–604.

Goldman, Y. E., Hibberd, M. G., & Trentham, D. R. (1984b). Initiation of active contraction by photogeneration of adenosine-5'-triphosphate in rabbit psoas muscle fibres. *Journal of Physiology, London, 354,* 605–24.

Goldman, Y. E., McCray, J. A., & Ranatunga, K. W. (1987). Transient tension changes initiated by laser temperature jumps in rabbit psoas muscle fibres. *Journal of Physiology, London, 392,* 71–95.

Goldman, Y. E., McCray, J. A., & Vallette, D. P. (1987). Cross-bridges in rigor fibres of rabbit psoas muscle support negative forces. *Journal of Physiology, London, 398,* 72P.

Goldman, Y. E., & Simmons, R. M. (1984). Control of sarcomere length in skinned muscle fibres of *Rana temporaria* during mechanical transients. *Journal of Physiology, London, 350,* 497–518.

Goldman, Y. E., & Simmons, R. M. (1986). The stiffness of frog skinned muscle fibres at altered lateral filament spacing. *Journal of Physiology, London, 378,* 175–94.

Gonzalez-Serratos, H. (1965). Inward spread of contraction during a twitch. *Journal of Physiology, London, 185,* 20–1P.

Gonzalez-Serratos, H. (1971). Inward spread of activation in vertebrate muscle fibres. *Journal of Physiology, London, 212,* 777–99.

Gonzalez-Serratos, H. (1975). Graded activation of myofibrils and the effect of diameter on tension development during contractures in isolated skeletal muscle fibres. *Journal of Physiology, London, 253,* 321–9.

Gonzalez-Serratos, H., Garcia, M. del C., Somlyo, A. V., Somlyo, A. P., & McClellan, G. (1981). Differential shortening of myofibrils during development of fatigue. *Biophysical Journal, 33,* 244a.

Gonzalez-Serratos, H., Somlyo, A. V., McClellan, G., Shuman, H., Borrero, L. M., & Somlyo, A. P. (1978). Composition of vacuoles and sarcoplasmic reticulum in fatigued muscle: electron probe analysis. *Proceedings of the National Academy of Sciences, USA, 75,* 1329–33.

Gordon, A. M., Huxley, A. F., & Julian, F. J. (1966a). Tension development in highly stretched vertebrate muscle fibres. *Journal of Physiology, London, 184,* 143–69.

Gordon, A. M., Huxley, A. F., & Julian, F. J. (1966b). The variation in isometric tension with sarcomere length in vertebrate muscle fibres. *Journal of Physiology, London, 184,* 170–92.

Gordon, A. M., & Ridgway, E. B. (1978). Calcium transients and relaxation in single muscle fibres. *European Journal of Cardiology, 7,* 27–34.

Gordon, A. M., & Ridgway, E. B. (1987). Extra calcium on shortening in barnacle muscle: is the decrease in calcium binding related to decreased cross-bridge attachment, force or length? *Journal of General Physiology, 90,* 321–40.

Gordon, A. M., & Ridgway, E. B. (1990). Stretch of active muscle during the declining phase of the calcium transient produces biphasic changes in calcium binding to the activating sites. *Journal of General Physiology, 96,* 1013–36.

Gordon, A. M., Ridgway, Yates, L. D., & Allen, T. (1988). Muscle cross-bridge attachment: effects on calcium binding and calcium activation. *Advances in Experimental Medicine and Biology, 226,* 89–98.

Grabarek, Z., Grabarek, J., Leavis, P. C., & Gergely, J. (1983). Cooperative binding to the Ca^{2+}-specific sites of troponin C in regulated actin and actomyosin. *Journal of Biological Chemistry, 258,* 14098–102.

Grabowski, W., Lobsiger, E. A., & Lüttgau, H. C. (1972). The effect of repetitive stimulation at low frequencies upon the electrical and mechanical activity of single muscle fibres. *Pflügers Archiv, 334,* 222–39.

Granzier, H. L. M., & Pollack, G. H. (1985). Stepwise shortening in unstimulated frog skeletal muscle fibres. *Journal of Physiology, London, 362,* 173–88.

Granzier, H. L. H., & Pollack, G. H. (1990). The descending limb of the force–sarcomere relation revisited. *Journal of Physiology, London, 421,* 595–615.

Greene, L. E. (1986). Cooperative binding of myosin subfragment one to regulated actin as measured by fluorescence changes of troponin I modified with different fluorophores. *Journal of Biological Chemistry, 261,* 1279–85.

Greene, L. E., & Eisenberg, E. (1980). Cooperative binding of myosin subfragment-1 to the actin–troponin tropomyosin complex. *Proceedings of the National Academy of Sciences, USA, 77,* 2616–20.

Gronert, G. A. (1980). Malignant hyperthermia. *Anesthesiology, 53,* 395–423.

Gulati, J., & Podolsky, R. J. (1981). Isotonic contraction of skinned muscle fibers on a slow time base. Effects of ionic strength and calcium. *Journal of General Physiology, 78,* 233–57.

Guth, K., & Potter, J. D. (1987). Effect of rigor and cycling cross-bridges on the structure of troponin C and on the Ca^{2+} affinity of the Ca^{2+}-specific regulatory sites in skinned rabbit psoas fibers. *Journal of Biological Chemistry, 262,* 13627–35.

Hall, Z. W., & Ralston, E. (1989). Nuclear domains in muscle cells. *Cell, 59,* 771–2.

Hamill, O. P., Marty, A., Neher, E., Sakmann, B., & Sigworth, F. J. (1981). Improved patch-clamp techniques for high-resolution current recording from cells and cell-free membrane patches. *Pflügers Archiv, 391,* 85–100.

Hanson, J., & Huxley, H. E. (1955). The structural basis of contraction in striated muscle. *Symposia of the Society for Experimental Biology, 9,* 228–64.

Harding, D. P., Kirschenlohr, H. L., Metcalfe, J. C., Morris, P. G., & Smith, G. A. (1989). The effect of stimulation frequency on end-diastolic $[Ca^{2+}]_i$ in isolated ferret heart. *Journal of Physiology, London, 417,* 51P.

Harrington, W. F. (1971). A mechanochemical mechanism for muscle contraction. *Proceedings of the National Academy of Sciences, USA, 68,* 685–9.

Harrington, W. F. (1979). On the origin of the contractile force in skeletal muscle. *Proceedings of the National Academy of Sciences, USA, 76,* 5066–70.

Harris, D. A., Falls, D. L., & Fischbach, G. D. (1989). Differential activation

274   References

of myotube nuclei following exposure to an acetylcholine receptor-inducing factor. *Nature, London, 337,* 173–6.

Harrison, G. G. (1975). Control of malignant hyperpyrexic syndrome in MHS swine by dantrolene sodium. *British Journal of Anesthesia, 47,* 62–5.

Haselgrove, J. C. (1973). X-ray evidence for a conformational change in the actin-containing filaments of vertebrate striated muscle. *Cold Spring Harbor Symposia on Quantitative Biology, 37,* 341–52.

Haselgrove, J. C. (1975). X-ray evidence for conformational changes in the myosin filaments of vertebrate striated muscle. *Journal of Molecular Biology, 92,* 113–43.

Haselgrove, J. C. (1983). Structure of vertebrate striated muscle as determined by x-ray diffraction studies. In *Handbook of Physiology, Section 10: Skeletal Muscle,* ed. L. D. Peachey, R. H. Adrian, & S. R. Geiger, pp. 143–71. Bethesda: American Physiological Society.

Haselgrove, J. C., & Huxley, H. E. (1973). X-ray evidence for radial cross-bridge movement and for the sliding filament model in actively contracting muscle. *Journal of Molecular Biology, 77,* 549–68.

Heistracher, P., & Hunt, C. C. (1969). The effect of procaine on snake twitch muscle fibres. *Journal of Physiology, London, 201,* 627–38.

Helland, L. A., Lopez, J. R., Taylor, S. R., Turbe, G., & Wanek, L. A. (1988). Effects of calcium "antagonists" on vertebrate skeletal muscle cells. *Proceedings of the New York Academy of Sciences, 522,* 259–68.

Herman, B., Nieminen, A. L., Gores, G. J., & Lemasters, J. J. (1988). Irreversible injury in anoxic hepatocytes precipitated by an abrupt increase in plasma membrane permeability. *FASEB Journal, 2,* 146–51.

Herring, B. P., & England, P. J. (1986). The turnover of phosphate bound to myosin light chain-2 in perfused rat heart. *Biochemical Journal, 240,* 205–14.

Hibberd, M. G., Dantzig, J. A., Trentham, D. R., & Goldman, Y. E. (1985). Phosphate release and force generation in skeletal muscle fibers. *Science, 228,* 1317–19.

Hibberd, M. G., & Trentham, D. R. (1986). Relationships between chemical and mechanical events during muscular contraction. *Annual Review of Biophysics and Biophysical Chemistry, 15,* 119–61.

Hilgemann, D. W., & Noble, D. (1987). Excitation–contraction coupling and extracellular calcium transients in rabbit atrium: reconstruction of basic cellular mechanisms. *Proceedings of the Royal Society [B], 230,* 163–205.

Hill, A. V. (1938). The heat of shortening and the dynamic constants of muscle. *Proceedings of the Royal Society of London [B], 126,* 136–95.

Hill, A. V. (1948). On the time required for diffusion and its relation to processes in muscle. *Proceedings of the Royal Society of London [B], 135,* 446–53.

Hill, A. V. (1949). The abrupt transition from rest to activity in muscle. *Proceedings of the Royal Society of London [B], 136,* 399–420.

Hill, D. K. (1964). The space accessible to albumin within the striated muscle fibre of the toad. *Journal of Physiology, London, 175*, 275–94.

Hill, L. (1977). A-band length, striation spacing and tension change on stretch of active muscle. *Journal of Physiology, London, 266*, 677–85.

Hill, T. L. (1974). Theoretical formalism for the sliding filament model of contraction of striated muscle. Part I. *Progress in Biophysics and Molecular Biology, 28*, 267–340.

Hill, T. L. (1975). Theoretical formalism for the sliding filament model of contraction of striated muscle. Part II. *Progress in Biophysics and Molecular Biology, 29*, 105–59.

Hill, T. L., Eisenberg, E., & Greene, L. (1980). Theoretical model for the cooperative equilibrium binding of myosin subfragment-1 to the actin–troponin–tropomyosin complex. *Proceedings of the National Academy of Sciences, USA, 77*, 3186–90.

Hill, T. L., Eisenberg, E., & Greene, L. (1983). Alternate model for the cooperative equilibrium binding of myosin subfragment-1–nucleotide complex to actin–troponin–tropomyosin. *Proceedings of the National Academy of Sciences, USA, 80*, 60–4.

Hille, B. (1977). Local anesthetics: hydrophilic and hydrophobic pathway for the drug receptor reaction. *Journal of General Physiology, 69*, 497–515.

Hille, B. (1989). Ionic channels: evolutionary origins and modern roles. *Quarterly Journal of Experimental Physiology, 74*, 785–804.

Hodgkin, A. L. (1958). The Croonian Lecture. Ionic movements and electrical activity in giant nerve fibres. *Proceedings of the Royal Society [B], 148*, 1–37.

Hodgkin, A. L. (1964). *The Conduction of the Nervous Impulse*. Liverpool: Liverpool University Press.

Hodgkin, A. L. (1977). Chance and design in electrophysiology: an informal account of certain experiments on nerve carried out between 1934 and 1952. In *The Pursuit of Nature*, by A. L. Hodgkin et al., pp. 1–21. Cambridge: Cambridge University Press.

Hodgkin, A. L., & Huxley, A. F. (1939). Action potentials recorded from inside a nerve fibre. *Nature, London, 144*, 710.

Hodgkin, A. L., & Huxley, A. F. (1945). Resting and action potentials in single nerve fibres. *Journal of Physiology, London, 104*, 176–95.

Hodgkin, A. L., & Huxley, A. F. (1946). Potassium leakage from an active nerve fibre. *Nature, London, 158*, 376.

Hodgkin, A. L., & Huxley, A. F. (1947a). Potassium leakage and absorption by an active nerve fibre. *Abstracts of Communications at the 17th International Congress of Physiological Sciences*, pp. 82–3.

Hodgkin, A. L., & Huxley, A. F. (1947b). Potassium leakage from an active nerve fibre. *Journal of Physiology, London, 106*, 341–67.

Hodgkin, A. L., & Huxley, A. F. (1951). Transport of radioactive potassium by current through a nerve membrane. *Journal of Physiology, London, 115,* 6P (title only).

Hodgkin, A. L., & Huxley, A. F. (1952a). Currents carried by sodium and potassium ions through the membrane of the giant axon *Loligo. Journal of Physiology, London, 116,* 449–72.

Hodgkin, A. L., & Huxley, A. F. (1952b). The components of membrane conductance in the giant axon of *Loligo. Journal of Physiology, London, 116,* 473–96.

Hodgkin, A. L., & Huxley, A. F. (1952c). The dual effect of membrane potential on sodium conductance in the giant axon of *Loligo. Journal of Physiology, London, 116,* 497–506.

Hodgkin, A. L., & Huxley, A. F. (1952d). A quantitative description of membrane current and its application to conduction and excitation in nerve. *Journal of Physiology, London, 117,* 500–44.

Hodgkin, A. L., & Huxley, A. F. (1952e). Propagation of electrical signals along giant nerve fibres. *Proceedings of the Royal Society [B], 140,* 177–83.

Hodgkin, A. L., & Huxley, A. F. (1952f). Movement of sodium and potassium ions during nervous activity. *Cold Spring Harbor Symposia on Quantitative Biology, 17,* 43–50.

Hodgkin, A. L., Huxley, A. F., & Katz, B. (1949). Ionic currents underlying activity in the giant axon of the squid. *Archives des Sciences Physiologiques, 3,* 129–50.

Hodgkin, A. L., Huxley, A. F., & Katz, B. (1952). Measurement of current-voltage relations in the membrane of the giant axon of *Loligo. Journal of Physiology, London, 116,* 424–48.

Hodgkin, A. L., & Nunn, B. J. (1987). The effects of ions on sodium-calcium exchange in salamander rods. *Journal of Physiology, London, 391,* 371–98.

Hofmann, P. A., & Fuchs, F. (1987). Evidence for a force-dependent component of calcium binding to cardiac troponin C. *American Journal of Physiology, 253,* C541–6.

Holroyde, M. E., Howe, E., & Solaro, J. (1979). Modification of calcium requirements for activation of cardiac myofibrillar ATPase by cAMP dependent phosphorylation. *Biochimica et Biophysica Acta, 587,* 628–37.

Homsher, E., & Lacktis, J. (1988). The effect of shortening on the phosphate release step of the actomyosin ATPase mechanism. *Biophysical Journal, 53,* 564a.

Horn, R., & Brodwick, M. S. (1980). Acetylcholine-induced current in perfused rat myoballs. *Journal of General Physiology, 75,* 297–321.

Horowits, R., Kempner, E. S., Bisher, M. E., & Podolsky, R. J. (1986). A physiological role for titin and nebulin in skeletal muscle. *Nature, London, 323,* 160–4.

Horowits, R., Maruyama, K., & Podolsky, R. J. (1989). Elastic behaviour of connectin filaments during thick filament movement in activated skeletal muscle. *Journal of Cell Biology, 109,* 2169–76.

Horowits, R., & Podolsky, R. J. (1987). The positional stability of thick filaments in activated skeletal muscle depends on sarcomere length: evidence for the role of titin filaments. *Journal of Cell Biology, 105,* 2217–23.

Horowits, R., & Podolsky, R. J. (1988). Thick filament movement and isometric tension in activated skeletal muscle. *Biophysical Journal, 54,* 165–71.

Housmans, P. R., Lee, N. K. M., & Blinks, J. R. (1983). Active shortening retards the decline of the intracellular calcium transient in mammalian heart muscle. *Science, 221,* 159–60.

Hume, J. R., & Uehara, A. (1986a). Properties of "creep currents" in single frog atrial cells. *Journal of General Physiology, 87,* 833–55.

Hume, J. R., & Uehara, A. (1986b). "Creep currents" in single frog atrial cells may be generated by electrogenic Na/Ca exchange. *Journal of General Physiology, 87,* 857–84.

Huxley, A. F. (1952). Applications of an interference microscope. *Journal of Physiology, London, 117,* 52–3P.

Huxley, A. F. (1954). A high-power interference microscope. *Journal of Physiology, London, 125,* 11–13P.

Huxley, A. F. (1957a). Muscle structure and theories of contraction. *Progress in Biophysics and Biophysical Chemistry, 7,* 255–318.

Huxley, A. F. (1957b). An ultramicrotome. *Journal of Physiology, London, 137,* 73–4P.

Huxley, A. F. (1959). Local activation of muscle. *Annals of the New York Academy of Sciences, 81,* 446–52.

Huxley, A. F. (1961). A micromanipulator. *Journal of Physiology, London, 157,* 5–7P.

Huxley, A. F. (1965). Shinkei dendō no butsurigaku. *Kagaku, 35,* 601–7.

Huxley, A. F. (1971). The activation of striated muscle and its mechanical response (The Croonian Lecture, 1967). *Proceedings of the Royal Society [B], 178,* 1–27.

Huxley, A. F. (1974). Muscular contraction (Review Lecture). *Journal of Physiology, London, 243,* 1–43.

Huxley, A. F. (1977). Looking back on muscle. In *The Pursuit of Nature,* by A. L. Hodgkin et al., pp. 23–64. Cambridge: Cambridge University Press.

Huxley, A. F. (1980). *Reflections on Muscle.* Liverpool: Liverpool Univeristy Press.

Huxley, A. F. (1984). Response to "Is stepwise shortening an artifact?" *Nature, London, 309,* 713–14.

Huxley, A. F. (1986). Comments on "Quantal mechanisms in cardiac contraction". *Circulation Research, 59,* 9–14.

Huxley, A. F. (1988). Prefatory chapter: muscular contraction. *Annual Review of Physiology, 50,* 1–16.

278   *References*

Huxley, A. F. (1990). A theoretical treatment of diffraction of light by a striated muscle fibre. *Proceedings of the Royal Society [B]*, *241*, 65–71.

Huxley, A. F., & Gordon, A. M. (1962). Striation patterns in active and passive shortening of muscle. *Nature, London, 193*, 280–1.

Huxley, A. F., Kearney, A., & Purvis, C. (1962). A dissecting microscope. *Journal of Physiology, London, 162*, 42–4P.

Huxley, A. F., & Lombardi, V. (1980). A sensitive force transducer with resonant frequency 50 kHz. *Journal of Physiology, London, 305*, 15–6P.

Huxley, A. F., Lombardi, V., & Peachey, L. D. (1981). A system for fast recording of longitudinal displacement of a striated muscle fibre. *Journal of Physiology, London, 317*, 12–13P.

Huxley, A. F., & Niedergerke, R. (1954). Interference microscopy of living muscle fibres. *Nature, London, 173*, 971–3.

Huxley, A. F., & Niedergerke, R. (1958). Measurement of the striations of isolated muscle fibres with the interference microscope. *Journal of Physiology, London, 144*, 403–25.

Huxley, A. F., & Peachey, L. D. (1961). The maximum length for contraction in vertebrate striated muscle. *Journal of Physiology, London, 156*, 150–65.

Huxley, A. F., & Peachey, L. D. (1964). Local activation of crab muscle. *Journal of Cell Biology, 23*, 107A.

Huxley, A. F., & Read, G. L. (1979). An automatic smoothing circuit for the input to digitizing equipment. *Journal of Physiology, London, 292*, 11–12P.

Huxley, A. F., & Simmons, R. M. (1968). A capacitance–gauge tension transducer. *Journal of Physiology, London, 197*, 12P.

Huxley, A. F., & Simmons, R. M. (1970). A quick phase in the series-elastic component of striated muscle, demonstrated in isolated fibres from the frog. *Journal of Physiology, London, 208*, 52–3.

Huxley, A. F., & Simmons, R. M. (1971a). Mechanical properties of the cross-bridges of frog striated muscle. *Journal of Physiology, London, 218*, 59–60P.

Huxley, A. F., & Simmons, R. M. (1971b). Proposed mechanism of force generation in striated muscle. *Nature, London, 233*, 533–8.

Huxley, A. F., & Simmons, R. M. (1973). Mechanical transients and the origin of muscular force. *Cold Spring Harbor Symposia on Quantitative Biology, 37*, 669–83.

Huxley, A. F., & Stämpfli, R. (1948). Beweis der saltatorischen Erregungsleitung im markhaltingen peripheren Nerven. *Helvetica Physiologica et Pharmacologica Acta, 6*, C22–4.

Huxley, A. F., & Stämpfli, R. (1949a). Evidence for saltatory conduction in peripheral myelinated nerve fibres. *Journal of Physiology, London, 108*, 315–39.

Huxley, A. F., & Stämpfli, R. (1949b). Saltatory transmission of the nervous impulse. *Archives des Sciences Physiologiques, 3*, 435–48.

Huxley, A. F., & Stämpfli, R. (1950a). Direct determination of the membrane potential of a myelinated nerve fibre at rest and in activity. *Abstracts of Communications at the 18th International Congress of Physiological Sciences*, 273–4.

Huxley, A. F., & Stämpfli, R. (1950b). Direkte Bestimmung des Membranpotentials der markhaltigen Nervenfaser in Ruhe und Erregung. *Helvetica Physiologica et Pharmacologica Acta, 8,* 107–9.

Huxley, A. F., & Stämpfli, R. (1951a). Direct determination of membrane resting potential and action potential in single myelinated nerve fibres. *Journal of Physiology, London, 112,* 476–95.

Huxley, A. F., & Stämpfli, R. (1951b). Effect of potassium and sodium on resting and action potentials of single myelinated nerve fibres. *Journal of Physiology, London, 112,* 496–508.

Huxley, A. F., & Straub, R. W. (1958). Local activation and interfibrillar structures in striated muscle. *Journal of Physiology, London, 143,* 40–1P.

Huxley, A. F., & Taylor, R. E. (1955a). Activation of a single sarcomere. *Journal of Physiology, London, 130,* 49–50P.

Huxley, A. F., & Taylor, R. E. (1955b). Function of Krause's membrane. *Nature, London, 176,* 1068.

Huxley, A. F., & Taylor, R. E. (1958). Local activation of striated muscle fibres. *Journal of Physiology, London, 144,* 426–41.

Huxley, H. E. (1953). Electron microscope studies of the organisation of the filaments in striated muscle. *Biochimica et Biophysica Acta, 12,* 387–94.

Huxley, H. E. (1957). The double array of filaments in cross-striated muscle. *Journal of Biophysical and Biochemical Cytology, 3,* 631–48.

Huxley, H. E. (1964). Evidence for continuity between the central elements of the triads and the extracellular space in frog sartorius muscle. *Nature, London, 202,* 1067–71.

Huxley, H. E. (1965). Structural evidence concerning the mechanism of contraction in striated muscle. In *Muscle*, ed. W. M. Paul, E. E. Daniel, C. M. Kay, & G. Monkton, pp. 3–28. Oxford: Pergamon Press.

Huxley H. E. (1969). The mechanism of muscular contraction. *Science, 164,* 1356–66.

Huxley, H. E. (1973). Structural changes in the actin- and myosin-containing filaments during contraction. *Cold Spring Harbor Symposia on Quantitative Biology, 37,* 361–76.

Huxley, H. E. (1979). Time resolved x-ray diffraction studies on muscle. In *Cross-Bridge Mechanism in Muscular Contraction*, ed. H. Sugi & G. H. Pollack, pp. 391–401. Tokyo: University of Tokyo Press.

Huxley, H. E., & Brown, W. (1967). The low-angle x-ray diagram of vertebrate striated muscle and its behaviour during contraction and rigor. *Journal of Molecular Biology, 30,* 383–434.

Huxley, H. E., Faruqi, A. R., Bordas, J., Koch, M. H. J., & Milch, J. R. (1980). The use of synchrotron radiation in time-resolved x-ray diffraction studies of myosin layer-line relections during muscle contraction. *Nature, London, 284,* 140–3.

Huxley, H. E., & Hanson, J. (1954). Changes in the cross-striations of muscle during contraction and stretch and their structural interpretation. *Nature, London, 173,* 973–6.

Huxley, H. E., & Kress, M. (1985). Crossbridge behaviour during muscle contraction. *Journal of Muscle Research and Cell Motility, 6,* 153–61.

Huxley, H. E., Kress, M., Faruqi, A. R., & Simmons, R. M. (1988). X-ray diffraction studies on muscle during rapid shortening and their implications concerning crossbridge behaviour. In *Molecular Mechanism of Molecular Contraction (Advances in Experimental Medicine and Biology,* Vol. 226), ed. H. Sugi & G. H. Pollack, pp. 347–52. New York: Plenum Press.

Huxley, H. E., Simmons, R. M., & Faruqi, A. R. (1989). Time course of spacing change of 143 meridional crossbridge reflection during rapid muscle shortening. *Biophysical Journal, 55,* 12a.

Huxley, H. E., Simmons, R. M., Faruqi, A. R., Kress, M., Bordas, J., & Koch, M. H. J. (1981). Millisecond time-resolved changes in x-ray reflections from contracting muscle during rapid mechanical transients, recorded using synchrotron radiation. *Proceedings of the National Academy of Sciences, USA, 78,* 2297–301.

Huxley, H. E., Simmons, R. M., Faruqi, A. R., Kress, M., Bordas, J., & Koch, M. H. J. (1983). Changes in the x-ray reflections from contracting muscle during rapid mechanical transients and their structural implications. *Journal of Molecular Biology, 169,* 469–506.

Hymel, L., Inui, M., Fleischer, S., & Schindler, H. G. (1988). Purified ryanodine receptor of skeletal muscle forms Ca^{2+}-activated oligomeric Ca^{2+} channels in planar bilayers. *Proceedings of the National Academy of Sciences, USA, 85,* 441–5.

Imagawa, T., Smith J. S., Coronado, R., & Campbell, K. P. (1987). Purified ryanodine receptor from skeletal muscle sarcoplasmic reticulum is the Ca^{2+}-permeable pore of the calcium release channel. *Journal of Biological Chemistry, 262,* 16636–43.

Infante, A. A., Klaupiks, D., & Davies, R. B. (1964). Adenosine triphosphate: changes in muscle doing negative work. *Science, 144,* 1577–8.

Inoue, S. (1986). *Video Microscopy.* New York: Plenum Press.

Inui, M., Saito, A., & Fleischer, S. (1987). Purification of the ryanodine receptor and identity with feet structures of junctional terminal cisternae of sarcoplasmic reticulum from fast skeletal muscle. *Journal of Biological Chemistry, 262,* 1740–7.

Isaacs, W. B., Kim, I. S., Struve, A., & Fulton, A. B. (1989). Biosynthesis of titin in cultured skeletal muscle cells. *Journal of Cell Biology, 109,* 2189–95.

Ishii, Y., & Lehrer, S. S. (1987). Fluorescence probe studies of the state of tropomyosin in reconstituted muscle thin filaments. *Biochemistry, 26,* 4922–5.

Ishizuka, T., & Endo, M. (1983). Effects of adenine on skinned fibers of amphibian fast skeletal muscle. *Proceedings of the Japan Academy, 59,* 93–6.

Ishizuka, T., Iijima, T., & Endo, M. (1983). Effects of adenine on twitch and other contractile responses of single fibers of amphibian fast skeletal muscle. *Proceedings of the Japan Academy, 59,* 97–100.

Jorgensen, A. O., Arnold, W., Pepper, D. R., Kahl, S. O., Mandel, F., & Campbell, K. P. (1988). A monoclonal antibody to the Ca^{2+}-ATPase of cardiac sarcoplasmic reticulum cross-reacts with slow type I but not with fast type II canine skeletal muscle fibers: an immunocytochemical and immunochemical study. *Cell Motility and Cytoskeleton, 9,* 164–74.

Julian, F. J. (1971). The effect of calcium on the force–velocity relation of briefly glycerinated frog muscle fibres. *Journal of Physiology, London, 218,* 117–45.

Julian, F. J., & Morgan, D. L. (1979a). Intersarcomere dynamics during fixed-end tetanic contractions of frog muscle fibres. *Journal of Physiology, London 293,* 365–78.

Julian, F. J., & Morgan, D. L. (1979b). The effect on tension of non-uniform distribution of length changes applied to frog muscle fibres. *Journal of Physiology, London, 293,* 379–92.

Julian, F. J., & Moss, R. L. (1981). Effects of calcium and ionic strength on shortening velocity and tension development in frog skinned muscle fibres. *Journal of Physiology, London, 311,* 179–99.

Julian, F. J., Sollins, K. R., & Sollins, M. R. (1974). A model for the transient and steady state mechanical behaviour of contracting muscle. *Biophysical Journal, 14,* 546–62.

Julian, F. J., & Sollins, M. R. (1975). Variation of muscle stiffness with force at increasing speeds of shortening. *Journal of General Physiology, 66,* 287–302.

Julian, F. J., Sollins, M. R., & Moss, R. L. (1978). Sarcomere length non-uniformity in relation to tetanic responses of stretched skeletal muscle fibres. *Proceedings of the Royal Society [B], 200,* 109–16.

Kalow, W., Britt, B. A., Terreau, M. E., & Haist, C. C. (1970). Metabolic error of muscle metabolism after recovery of malignant hyperthermia. *Lancet, II,* 895–8.

Kaplan, J. H., Forbush, B., III, & Hoffman, J. F. (1978). Rapid photolytic release of adenosine 5'-triphosphate from a protected analogue: utilization by the Na:K pump of human red blood cell ghosts. *Biochemistry, 17,* 1929–35.

Kato, N. S., Weisberg, A., & Winegrad, S. (1991). Effect of left atrial pressure on the activity of specific myosin isozymes in rat heart. *Circulation Research, 68,* 1582–90.

Katz, B. (1939). The relation between force and speed in muscular contraction. *Journal of Physiology, London, 96,* 45–64.

Kawamoto, R. M., Brunschwig, J. P., & Caswell, A. H. (1988). Localization by immunoelectron microscopy of spanning protein of triad junction in terminal cisternae/triad vesicles. *Journal of Muscle Research and Cell Motility, 9*, 334–43.

Kawamoto, R. M., Brunschwig, J. P., Kim, K. C., & Caswell, A. H. (1986). Isolation, characterization, and localization of the spanning protein from skeletal muscle triads. *Journal of Cell Biology, 103*, 1405–14.

Kelly, A. M. (1980). T tubules in neonatal rat soleus and extensor digitorum longus muscle. *Developmental Biology, 80*, 501–5.

Kimura, J., Miyamae, S., & Noma, A. (1987). Identification of sodium–calcium exchange current in single ventricular cells of guinea-pig. *Journal of Physiology, London, 384*, 199–222.

Knudson, C. M., Chaudhari, N., Sharp, A. H., Powell, J. A., Beam, K. G., & Campbell, K. P. (1989). Specific absence of the alpha-1 subunit of the dihydrophyridine receptor in mice with muscular dysgenesis. *Journal of Biological Chemistry, 264*, 1345–8.

Kobayashi, T., & Endo. M. (1988). Temperature-dependent inhibition of caffeine contracture of mammalian skeletal muscle by dantrolene. *Proceedings of the Japan Academy, 64*, 76–9.

Koch-Weser, J., & Blinks, J. R. (1963). The influence of the interval between beats on myocardial contractility. *Pharmacological Reviews, 15*, 601–52.

Kolb, M. E., Horne, M. L., & Martz, R. (1982). Dantrolene in human malignant hyperthermia. *Anesthesiology, 56*, 254–62.

Kress, M., Huxley, H. E., Faruqi, A. R., & Hendrix, J. (1986). Structural changes during activation of frog muscle studied by time-resolved x-ray diffraction. *Journal of Molecular Biology, 188*, 325–42.

Krueger, J. W., Forletti, D., & Wittenberg, B. A. (1980). Uniform sarcomere shortening behavior in isolated cardiac muscle cells. *Journal of General Physiology, 76*, 587–607.

Kushmerick, M. J., & Davies, R. E. (1969). The chemical energetics of muscle contraction. II. The chemistry, efficiency and power of maximally working sartorius muscles. *Proceedings of the Royal Society [B], 174*, 315–53.

Lado, M. G., Sheu, S.-S., & Fozzard, H. A. (1982). Changes in intracellular Ca^{2+} activity with stimulation in sheep cardiac Purkinje strands. *American Journal of Physiology, 243*, H133–7.

Lai, F. A., Erickson, H. P., Rousseau, E., Liu, Q-Y., & Meissner, G. (1988). Purification and reconstitution of the calcium release channel from skeletal muscle. *Nature, London, 331*, 315–9.

Lännergren, J. (1977). Location of U.V.-absorbing substance in isolated skeletal muscle fibres. The effect of stimulation. *Journal of Physiology, London, 270*, 785–800.

Lännergren, J. (1978). The force–velocity relation of isolated twitch and slow muscle fibres of *Xenopus laevis. Journal of Physiology, London, 283*, 501–21.

Lännergren, J. (1979). An intermediate type of muscle fibre in *Xenopus laevis. Nature, London, 279*, 254–6.

Lännergren J. (1987). Contractile properties and myosin isoenzymes of various kinds of *Xenopus* twitch muscle fibres. *Journal of Muscle Research and Cell Motility, 8,* 260–73.

Lännergren, J., & Hoh, J. F. Y. (1984). Myosin isoenzymes in single muscle fibres of *Xenopus laevis,* analysis of five different functional types. *Proceedings of the Royal Society [B], 222,* 401–8.

Lännergren, J., Lindblom, P., & Johansson, B. (1982). Contractile properties of two varieties of twitch muscle fibres in *Xenopus laevis. Acta Physiologica Scandinavica, 114,* 523–35.

Lännergren, J., & Smith, R. S. (1966). Types of muscle fibres in toad skeletal muscle. *Acta Physiologica Scandinavica, 68,* 263–74.

Lännergren, J., & Westerblad, H. (1987). Action potential fatigue in single skeletal muscle fibres of *Xenopus. Acta Physiologica Scandinavica, 129,* 311–18.

Lännergren, J., & Westerblad, H. (1989). Maximum tension and force–velocity properties of fatigued, single *Xenopus* muscle fibres studied by caffeine and high K^+. *Journal of Physiology, London, 409,* 473–90.

Lass, Y., & Fischbach, G. D. (1976). A discontinuous relationship between the acetylcholine-activated channel conductance and temperature. *Nature, London, 263,* 150–1.

Lazarides, E., & Capetanaki, Y. G. (1986). The striated muscle cytoskeleton: expression and assembly in development. In *Molecular Biology of Muscle Development,* ed. C. Emerson et al., pp. 749–72. New York: Alan R. Liss.

Leavis, P. C., & Gergely, J. (1984). Thin filament proteins and thin filament-linked regulation of vertebrate muscle contraction. *CRC Critical Reviews in Biochemistry, 16,* 235–305.

Lee, J. A., Westerblad, H. & Allen, D. G. (1991). Changes in tetanic and resting Ca^{2+} during fatigue and recovery of single muscle fibres from *Xenopus laevis. Journal of Physiology, London, 433,* 307–26.

Lehrer, S. S., & Morris, E. D. (1982). Dual effects of tropomyosin and troponin–tropomyosin on actomyosin subfragment 1 ATPase. *Journal of Biological Chemistry, 257,* 8073–80.

Leung, A. T., Imagawa, T., Block, B., Franzini-Armstrong, C., & Campbell, K. P. (1988). Biochemical and ultrastructural characterization of the 1,4-dihydropyridine receptor from rabbit skeletal muscle. Evidence for a 52,000-Da subunit. *Journal of Biological Chemistry, 263,* 994–1001.

Lieb, H., & Loewi, O. (1918). Über Spontanerholung des Froscheherzens bei unzureichender Kationenspeisung. III Mitteilung. Quantitative mikroanalytische Untersuchungen über die Ursache der Calciumabgabe von seiten des Herzens. *Pflügers Archiv, 173,* 152–7.

Lin, T. I., Jovanovic, M. V., & Dowben, R. M. (1989). Nine new fluorescent probes. *Society of Photo-optical Instrumentation Engineers Symposium, Fluorescent Probe Molecules: Innovations and Applications, 1063,* 133–41.

Loesser, K. E., Castellani, L., & Franzini-Armstrong, C. (1992). Disposition of junctional feet in muscles of invertebrates. *Journal of Muscle Research and Cell Motility, 13*.

Lombardi, V., & Piazzesi, G. (1990). The contractile response during steady lengthening of stimulated frog muscle fibres. *Journal of Physiology, London, 431*, 141–71.

Lymn, R. W., & Taylor, E. W. (1971). Mechanism of adenosine triphosphate hydrolysis by actomyosin. *Biochemistry, 10*, 4617–24.

MacLennan, D. H., & Wong, P. T. S. (1971). Isolation of a calcium-sequestering protein from sarcoplasmic reticulum. *Proceedings of the National Academy of Sciences, U.S.A., 68*, 1231–5.

Mainwood, G. W., & Renaud, J. M. (1985). The effect of acid–base balance on fatigue of skeletal muscle. *Canadian Journal of Physiology and Pharmacology, 63*, 403–16.

Mandelkow, E., Mandelkow, E.-M., Hotani, H., Hess, B., & Müller, S. C. (1989). Spatial patterns from oscillating microtubules. *Science, 246*, 1291–3.

Marban, E., Kitakaze, M., Kusuoka, H., Porterfield, J. K., Yue, D. T., & Chacko, V. P. (1987). Intracellular free calcium concentration measured with ^{19}F NMR spectroscopy in intact ferret hearts. *Proceedings of the National Academy of Sciences, USA, 84*, 6005–9.

Maruyama, K. (1986). Connectin, an elastic filamentous protein of striated muscle. *International Review of Cytology, 104*, 81–114.

Matsubara, I. (1980). X-ray diffraction studies of the heart. *Annual Review of Biophysics and Bioengineering, 9*, 81–105.

Matsubara, I., & Elliott, G. F. (1972). X-ray diffraction studies on skinned single fibres of frog skeletal muscle. *Journal of Molecular Biology, 72*, 657–69.

Matsubara, I., Goldman, Y. E., & Simmons, R. M. (1984). Changes in the lateral filament spacing of skinned muscle fibres when cross-bridges attach. *Journal of Molecular Biology, 173*, 15–33.

Matsubara, I., Maughan, D. W., Saeki, Y., & Yagi, N. (1989). Cross-bridge movement in rat cardiac muscle as a function of calcium concentration. *Journal of Physiology, London, 417*, 555–65.

Matsubara, I., & Yagi, N. (1985). Movements of cross-bridges during and after slow length changes in active frog skeletal muscle. *Journal of Physiology, London, 361*, 151–63.

Matsubara, I., Yagi, N., & Endoh, M. (1979). Movement of myosin heads during a heart beat. *Nature, London, 278*, 474–6.

Matsuda, H., & Stanfield, P. R. (1989). Single inwardly rectifying potassium channels in cultured muscle cells from rat and mouse. *Journal of Physiology, London, 414*, 111–24.

Matsui, K., & Endo, M. (1986). Effect of inhalation anesthetics on the rate of Ca release from the sarcoplasmic reticulum of skeletal muscle in the guinea pig. *Japanese Journal of Pharmacology, 40*, 245P.

McClellan, G., & Winegrad, S. (1978). The regulation of calcium sensitivity of the contractile system in mammalian cardiac muscle. *Journal of General Physiology, 72,* 737–64.

Meeder, T., & Ulbricht, W. (1987). Action of benzocaine on sodium channels of frog nodes of Ranvier treated with chloramine-T. *Pflügers Archiv, 409,* 265–73.

Meissner, G. (1975). Isolation and characterization of two types of sarcoplasmic reticulum vesicles. *Biochimica et Biophysica Acta, 389,* 51–68.

Meissner, G. (1986). Ryanodine activation and inhibition of the Ca^{2+} release channel of sarcoplasmic reticulum. *Journal of Biological Chemistry, 261,* 6300–6.

Meissner, G., Darling, E., & Eveleth, J. (1986). Kinetics of rapid Ca^{2+} release by sarcoplasmic reticulum: effects of Ca^{2+}, Mg^{2+}, and adenine nucleotides. *Biochemistry, 25,* 236–44.

Meltzer, W., Rios, E., & Schneider, M. F. (1984). Time course of calcium release and removal in skeletal muscle fibers. *Biophysical Journal, 45,* 637–41.

Meltzer, W., Rios, E., & Schneider, M. F. (1986). The removal of myoplasmic free calcium following calcium release in frog skeletal muscle. *Journal of Physiology, London, 372,* 261–92.

Metzger, J. M., Greaser, M. L., & Moss, R. L. (1989). Variations in cross-bridge attachment rate and tension with phosphorylation of myosin in mammalian skinned skeletal fibers. *Journal of General Physiology, 93,* 885–83.

Milligan, R. A., & Flicker, P. F. (1987). Structural relationships of actin, myosin and tropomyosin revealed by cryo-electron microscopy. *Journal of Cell Biology, 105,* 29–39.

Miura, Y., & Kirura, J. (1989). Sodium–calcium exchange current: dependence on internal Ca and Na and competitive binding of external Na and Ca. *Journal of General Physiology, 93,* 1129–45.

Mope, L., McClellan, G. G., & Winegrad, S. (1980). Calcium sensitivity of the contractile system and phosphorylation of troponin in hyperpermeable cardiac cells. *Journal of General Physiology, 75,* 271–82.

Morgan, D. L. (1990). New insights into the behaviour of muscle during active lengthening. *Biophysical Journal, 57,* 209–21.

Morgan, J. P., & Blinks, J. R. (1982). Intracellular Ca^{++} transients in cat papillary muscle. *Canadian Journal of Physiology and Pharmacology, 60,* 524–8.

Morris, V. A., Lopez, J. R., Parra, L., Rolli, M. L., & Taylor, S. R. (1989). Stimulus–contraction coupling in rat skeletal myoballs studied by computer vision. *The Physiologist, 32,* 17.

Moss, R. L. (1986). Effects on shortening velocity of rabbit skeletal muscle due to variations in the level of thin filament activation. *Journal of Physiology, London, 377,* 487–505.

Moss, R. L., Giulian, G. G., & Greaser, M. L. (1986). Effects of partial extraction of troponin complex upon the tension–pCa relation in rabbit skeletal muscle. *Journal of General Physiology, 87,* 761–74.

Mullins, L. J. (1977). A mechanism for Na/Ca transport. *Journal of General Physiology, 70,* 681–95.

Nassar-Gentina, V., Passonneau, J. V., & Rapoport, S. I. (1981). Fatigue and metabolism of frog muscle fibers during stimulation and in response to caffeine. *American Journal of Physiology, 241,* 160–6.

Needham, D. M. (1950). Myosin and adenosine triphosphate in relation to muscle contraction. *Biochimica et Biophysica Acta, 4,* 42–9.

Nelson, T. (1984). Dantrolene does not block calcium pulse-induced calcium release from a putative calcium channel in sarcoplasmic reticulum from malignant hyperthermia and normal pig muscle. *FEBS Letters, 167,* 123–6.

Nelson, W. J., & Lazarides, E. (1984). Assembly and establishment of membrane–cytoskeleton domains during differentiation. In *Cell Membranes: Methods and Reviews,* ed. E. Elias, W. Frazier, & L. Glaser, Vol. 2, pp. 219–46. New York: Plenum Press.

Newham, D. J., McPhail, G., Mills, K. R., & Edwards, R. H. T. (1983). Ultrastructural changes after concentric and eccentric contractions of human muscle. *Journal of the Neurological Sciences, 61,* 109–22.

Nicotera, P., Thor, H., & Orrenius, S. (1989). Cytosolic-free Ca^{2+} and cell killing in hepatoma lclc7 cells exposed to chemical anoxia. *FASEB Journal, 3,* 59–64.

Niedergerke, R. (1955). Local muscular shortening by intracellularly applied calcium. *Journal of Physiology, London, 128,* 12–3P.

Niedergerke, R. (1956). The 'staircase' phenomenon and the action of calcium on the heart. *Journal of Physiology, London, 134,* 569–83.

Niedergerke, R., Ogden, D. C., & Page, S. (1976). Contractile activation and calcium movements in heart cells. In *Calcium in Biological Systems, Symposium of the Society for Experimental Biology, 30,* 381–95.

Niedergerke, R., & Page, S. (1977). Analysis of catecholamine effects in single atrial trabeculae of the frog heart. *Proceedings of the Royal Society [B], 197,* 333–62.

Niedergerke, R., & Page, S. (1979). The hypodynamic condition of the frog heart. *Naunyn-Schmiedeberg's Archives of Pharmacology, 308,* R5.

Niedergerke, R. & Page, S. (1981). Analysis of caffeine action in single trabeculae of the frog heart. *Proceedings of the Royal Society [B], 213,* 303–24.

Niedergerke, R., & Page, S. (1989). Receptor-controlled calcium discharge in frog heart cells. *Quarterly Journal of Experimental Physiology, 74,* 987–1002.

Niedergerke, R., & Page, S., & Talbot, M. S. (1969). Determination of calcium movements in heart ventricles of the frog. *Journal of Physiology, London, 202,* 58–60P.

Noble, D. (1966). Application of Hodgkin-Huxley equations to excitable tissues. *Physiological Reviews, 46,* 1–50.

Noble, S., & Shimoni, Y. (1981). The calcium and frequency dependence of the slow inward current in frog atrium. *Journal of Physiology, London, 310,* 57–75.

Noda, M., Shepherd, R. N. & Gadsby, D. C. (1988). Activation by $[Ca]_i$, and block by 3',4'-dichlorobenzamil, of outward Na/Ca exchange current in guinea-pig ventricular myocytes. *Biophysical Journal, 53,* 342a.

O'Brien, E. J., Bennett, P. M., & Hanson, J. (1971). Optical diffraction studies of myofibrillar structure. *Philosophical Transactions of the Royal Society [B], 261,* 201–8.

Ohnishi, S. T., Taylor, S., & Gronert, G. A. (1983). Calcium-induced Ca^{2+} release from sarcoplasmic reticulum of pigs susceptible to malignant hyperthermia. *FEBS Letters, 161,* 103–7.

Ohta, T., & Endo, M. (1986). Inhibition of calcium-induced calcium release by dantrolene at mammalian body temperature. *Proceedings of the Japan Academy, 62,* 329–32.

Ohta, T., Endo, M., Nakano, K., Morohoshi, Y., Wanikawa, K., & Ohga, A. (1989). Ca-induced Ca release in malignant hyperthermia-susceptible pig skeletal muscle. *American Journal of Physiology, 256,* C358–C367.

Ohtsuki, I., Maruyama, K., & Ebashi, S. (1986). Regulatory and cytoskeletal proteins of vertebrate skeletal muscle. *Advances in Protein Chemistry, 38,* 1–67.

Oxford, G. S., & Pooler, J. P. (1975). Selective modification of sodium channel gating in lobster axons by 2,4,6-trinitrophenol. Evidence for two inactivation mechanisms. *Journal of General Physiology, 66,* 765–79.

Oyamada, H., Iino, M., & Endo, M. (1988). Open lock by ryanodine of the Ca-induced Ca release channels in skinned skeletal muscle of fibers requires the channel activation. *Japanese Journal of Pharmacology, 46,* 312P.

Page, S. (1964). The organisation of the sarcoplasmic reticulum in frog muscle. *Journal of Physiology, London, 175,* 10–11P.

Page, S. G. (1968). Fine structure of tortoise skeletal muscle. *Journal of Physiology, London, 197,* 709–15.

Page, S. G., & Huxley, H. E. (1963). Filament lengths in striated muscle. *Journal of Cell Biology, 19,* 369–90.

Page, S. G., & Niedergerke, R. (1972). Structures of physiological interest in the frog heart ventricle. *Journal of Cell Science, 11,* 179–203.

Parry, D. A., & Squire, J. M. (1973). The structual role of tropomyosin in muscle regulation, analysis of the x-ray diffraction patterns from relaxed and contracting muscles. *Journal of Molecular Biology, 75,* 33–55.

Partridge, L. D. (1966). Signal-handling characteristics of load-moving skeletal muscle. *American Journal of Physiology, 210,* 1178–91.

Peachey, L. D. (1967). Membrane systems in crab fibers. *American Zoology, 7,* 505–13.

Peachey, L. D., & Huxley, A. F. (1960). Local activation and structure of slow striated muscle fibers of the frog. *Federation Proceedings, 19,* 257.

Peachey, L. D., & Huxley, A. F. (1964). Transverse tubules in crab muscle. *Journal of Cell Biology, 23,* 70–1a.

Peganov, E. M., Khodorov, B. I., & Shiskova, L. D. (1973). Slow sodium inactivation in the Ranvier node membrane: role of external potassium. *Bulletin of Experimental Biology and Medicine, USSR, 76*, 1014–17.

Podolin, R. A., & Ford, L. E. (1983). The influence of calcium on the shortening velocity of skinned frog muscle cells. *Journal of Muscle Research and Cell Motility, 4*, 263–82.

Podolin, R. A., & Ford, L. E. (1986). The influence of partial activation on the force–velocity properties of frog skinned muscle in mM magnesium ion. *Journal of General Physiology, 87*, 607–31.

Podolsky, R. J. (1960). Kinetics of muscular contraction: the approach to the steady state. *Nature, London, 188*, 666–8.

Podolsky, R. J. (1962). Mechanochemical basis of muscular contraction. *Federation Proceedings, 21*, 964–74.

Podolsky, R. J., & Nolan, A. C. (1971). Cross-bridge properties derived from physiological studies of frog muscle fibers. In *Contractility of Muscle Cells*, ed. R. J. Podolsky, pp. 247–60. Englewood Cliffs, N.J.: Prentice-Hall.

Podolsky, R. J., Nolan, A. C., & Zaveler, S. A. (1969). Cross-bridge properties derived from muscle isotonic velocity transients. *Proceedings of the National Academy of Sciences, USA, 64*, 504–11.

Podolsky, R. J., St. Onge, R., Yu, L., & Lymn, R. W. (1976). X-ray diffraction of actively shortening muscle. *Proceedings of the National Academy of Sciences, USA, 73*, 813–17.

Podolsky, R. J., & Teichholz, L. E. (1970). The relation between calcium and contraction kinetics in skinned muscle fibres. *Journal of Physiology, London, 211*, 19–35.

Pollack, G. H. (1983). The cross-bridge theory. *Physiological Reviews, 63*, 1049–113.

Pollack, G. H., Iwazumi, T., ter Keurs, H. E. D. J., & Shibata, E. F. (1977). Sarcomere shortening in striated muscle occurs in stepwise fashion. *Nature, London, 268*, 757–9.

Porter, K. R., & Palade, G. E. (1957). Studies on the endoplasmic reticulum. *Journal of Biophysical and Biochemical Cytology, 3*, 269–300.

Porter, R., & Whelan, J. (1981). *Human Muscle Fatigue: Physiological Mechanisms. Ciba Foundation Symposium, 82*, ed. R. Porter & J. Whelan. London: Pitman Medical.

Ralston, E., & Hall, Z. W. (1989). Transfer of a protein encoded by a single nucleus to nearby nuclei in multinucleated myotubes. *Science, 244*, 1066–9.

Ramsey, R. W., & Street, S. F. (1940). The isometric length–tension diagram of isolated skeletal muscle fibers of the frog. *Journal of Cellular and Comparative Physiology, 15*, 11–34.

Rasgado-Flores, H., Santiago, E. M., & Blaustein, M. P. (1989). Kinetics and stoichiometry of coupled Na efflux and Ca influx (Na/Ca exchange) in barnacle muscle cells. *Journal of General Physiology, 93*, 1219–41.

Reeves, J. P., & Philipson, K. D. (1989). Sodium–calcium exchange activity in plasma membrane vesicles. In *Sodium–Calcium Exchange*, ed. T. J. A. Allen, D. Noble, & H. Reuter, pp. 27–53. Oxford: Oxford University Press.

Reuben, J. P., Brandt, P. W., Berman, M., & Grundfest, H. (1971). Regulation of tension in the skinned crayfish muscle fiber. I. Contraction and relaxation in the absence of Ca (pCa>9). *Journal of General Physiology, 57*, 385–407.

Ricker, K., Camacho, L. M., Grafe, P., Lehmann-Horn, F., & Rüdel, R. (1989). Adynamia episodica hereditaria: what causes the weakness? *Muscle and Nerve, 12*, 883–91.

Ridgway, E. B., & Gordon, A. M. (1984). Muscle calcium transient: effect of post-stimulus length changes in single muscle fibers. *Journal of General Physiology, 83*, 75–103.

Ridgway, E. B., Gordon, A. M., & Martyn, D. M. (1983). Hysteresis in the force–calcium relation in muscle. *Science, 219*, 1075–7.

Ringer, S. (1883). A further contribution regarding the influence of the different constituents of the blood on the contraction of the heart. *Journal of Physiology, London, 4*, 29–47.

Rios, E., & Brum, G. (1987). Involvement of dihydropyridine receptors in excitation–contraction coupling in skeletal muscle. *Nature, London, 325*, 717–20.

Rios, E., & Pizarro, G. (1988). Voltage sensors and calcium channels of excitation–contraction coupling. *News in Physiological Science, 3*, 223–7.

Robertson, J. D. (1956). Some features of the ultrastructure of reptilian skeletal muscle. *Journal of Biophysical and Biochemical Cytology, 2*, 369–80.

Robertson, S. P., Johnson, J. D., & Potter, J. D. (1981). The time-course of Ca^{2+} exchange with calmodulin, troponin, parvalbumin, and myosin in response to transient increases in Ca^{2+}. *Biophysical Journal, 34*, 559–69.

Rolli, M. L., Wanek, L. A., & Taylor, S. R. (1988). Contractile dynamics of rat skeletal myocytes measured by high-speed digital imaging microscopy. *Biophysical Journal, 53*, 172a.

Roos, K. P., Bliton, A. C., Lubell, B. A., Parker, J. M., Patton, M. J., & Taylor, S. R. (1989a). High speed striation pattern recognition in contracting cardiac myocytes. In *Proceedings of the Society of Photo-optical Instrumentation Engineers, New Technologies in Cytometry*, ed. G. C. Salzman, *1063*, 29–41.

Roos, K. P., Bliton, A. C., Patton, M. J., & Taylor, S. R. (1989b). Regional differences during unloaded contractions of isolated rat cardiac myocytes. *Biophysical Journal, 55*, 268a.

Roos, K. P., & Brady, A. J. (1982). Individual sarcomere length determination from isolated cardiac cells using high resolution optical microscopy and digital image processing. *Biophysical Journal, 40*, 233–44.

Roos, K. P., & Parker, J. M. (1990). A low cost two-dimensional digital image acquisition sub-system for high speed microscopic motion detection. In *Pro-*

ceedings of the Society of Photo-optical Instrumentation Engineers: Bioimaging and Two-Dimensional Spectroscopy, ed. L. C. Smith, *1205*, 134–41.

Roos, K. P., & Taylor, S. R. (1989). Striation dynamics in isolated rat cardiac myocytes revealed by computer vision. *Journal of Physiology, London, 418*, 50P.

Rousseau, E., LaDine, J., Liu, Q.-Y., & Meissner, G. (1988). Activation of the Ca^{2+} release channel of skeletal muscle sarcoplasmic reticulum by caffeine and related compounds. *Archives of Biochemistry and Biophysics, 267*, 75–86.

Rowlerson, A. M., & Spurway, N. C. (1988). Histochemical and immunohistochemical properties of skeletal muscle fibres from *Rana* and *Xenopus*. *Histochemical Journal, 20*, 657–73.

Rudy, B. (1978). Slow inactivation of the sodium conductance in squid giant axons: pronase resistance. *Journal of Physiology, London, 283*, 1–21.

Rüdel, R., Dengler, R., Ricker, K., Haass, A., & Emser, W. (1980). Improved therapy of myotonia with the lidocaine derivative tocainide. *Journal of Neurology, 222*, 275–78.

Rüdel, R., & Lehmann-Horn, F. (1985). Membrane changes in cells from myotonia patients. *Physiological Reviews, 63*, 310–56.

Rüdel, R. & Ricker, K. (1985). The primary periodic paralyses. *Trends in Neurosciences, 8*, 467–70.

Rüdel R., Ricker, K., & Lehmann-Horn, F. (1988). Transient weakness and altered membrane characteristic in recessive generalized myotonia (Becker). *Muscle and Nerve, 11*, 202–11.

Rüdel, R., & Taylor, S. R. (1971). Striated muscle fibers: facilitation of contraction at short lengths by caffeine. *Science, 172*, 387–8.

Rüdel, R., & Zite-Ferenczy, F. (1979). Do laser diffraction studies on striated muscle indicate stepwise sarcomere shortening? *Nature, London, 278*, 573–5.

Rüdel, R., & Zite-Ferenczy, F. (1980). Efficiency of light diffraction by cross-striated muscle fibers under stretch and during isometric contraction. *Biophysical Journal, 30*, 507–16.

Ruff, R. L., Simoncini, L., & Stühmer, W. (1988). Slow sodium channel inactivation in mammalian muscle: a possible role in regulating excitability. *Muscle and Nerve, 11*, 502–10.

Ruppersberg, J. P., Schure, A., & Rüdel, R. (1987). Inactivation of TTX-sensitive and TTX-insensitive sodium channels of rat myoballs. *Neuroscience Letters, 78*, 166–70.

Saito, A., Inui, M., Radermacher, M., Frank, J., & Fleischer, S. (1988). Ultrastructure of the calcium release channel of sarcoplasmic reticulum. *Journal of Cell Biology, 107*, 211–19.

Schauf, C. L., Pencek, T. L., & Davis, F. A. (1976). Slow sodium inactivation in *Myxicola* axons. *Biophysical Journal, 16*, 771–8.

Schiaffino, S., Cantini, M., & Sartore, S. (1977). T-system formation in cultured rat skeletal tissue. *Tissue and Cell, 9*, 437–46.

Schneider, M. F., & Chandler, W. K. (1973). Voltage dependent charge movement in skeletal muscle: a possible step in excitation–contraction coupling. *Nature, London, 242,* 244–6.

Schoenberg, M. (1989). Effect of adenosine triphosphate analogues on skeletal muscle fibers in rigor. *Biophysical Journal, 56,* 33–41.

Sheu, S.-S., & Fozzard, H. A. (1982). Transmembrane Na^+ and Ca^{2+} electrochemical gradients in cardiac muscle and their relationship to force development. *Journal of General Physiology, 80,* 325–51.

Sigworth, F. J., & Sine, S. M. (1987). Data transformation for improved display and fitting of single-channel dwell time histograms. *Biophysical Journal, 52,* 1047–54.

Silver, P., Buja, L. M., & Stull, J. (1986). Frequency-dependent myosin light chain phosphorylation in isolated myocardium. *Journal of Molecular and Cellular Cardiology, 18,* 31–7.

Simmons, R. M., & Jewell, B. R. (1974). Mechanics and models of muscular contraction. *Recent Advances in Physiology, 9,* 87–147.

Simoncini, L., & Stühmer, W. (1987). Slow sodium channel inactivation in rat fast twitch muscle. *Journal of Physiology, London, 383,* 327–37.

Smith, D. S. (1961). The structure of insect fibrillar flight muscle. *Journal of Biophysical and Biochemical Cytology, 10* (Suppl.), 123–58.

Smith, J. S., Coronado, R., & Meissner, G. (1986). Single channel measurements of the calcium release channel from skeletal muscle sarcoplasmic reticulum. Activation by Ca^{2+} and ATP and modulation by Mg^{2+}. *Journal of General Physiology, 88,* 573–88.

Smith, J. S., Imagawa, T., Ma, J., Fill, M., Campbell, K. P., & Coronado, R. (1988). Purified ryanodine receptor from rabbit skeletal muscle is the calcium-release channel of sarcoplasmic reticulum. *Journal of General Physiology, 92,* 1–26.

Smith, R. S., & Ovalle, W. K. (1973). Varieties of fast and slow extrafusal muscle fibres in amphibian hind limb muscles. *Journal of Anatomy, 116,* 1–24.

Somlyo, A. P. (1986). Recent advances in electron and light optical imaging in biology and medicine. *Annals of the New York Academy of Sciences, 483,* 1–469.

Somlyo, A. V., Gonzalez-Serratos, H., McClellan, G., Shuman, H., Borrero, L. M., & Somlyo, A. P. (1978). Electron probe analysis of the sarcoplasmic reticulum and vacuolated T-tubule system of fatigued frog muscles. *Annals of the New York Academy of Sciences, 307,* 232–4.

Somlyo, A. V., Gonzalez-Serratos, H., Shuman, H., McClellan, G., & Somlyo, A. P. (1981). Calcium release and ionic changes in the sarcoplasmic reticulum of tetanized muscle; an electron-probe study. *Journal of Cell Biology, 90,* 577–94.

Spande, J. I., & Schottelius, B. A. (1970). Chemical basis of fatigue in isolated mouse soleus muscle. *American Journal of Physiology, 219,* 1490–5.

Spurway, N. C. (1985). Positive correlation between oxidative and glycolytic capacities in frog muscle fibres. *IRCS Medical Science, 13,* 78–9.

Squire, J. (1981). *The Structural Basis of Muscular Contraction.* New York: Plenum Press.

Squire, J. M., Luther, P. K., & Morris, E. P. (1990). Organisation and properties of the striated muscle sarcomere. In *Molecular Mechanisms in Muscular Contraction,* ed. J. M. Squire, pp. 1–37. London: Macmillan Press.

Street, S. F. (1983). Lateral transmission of tension in frog myofibers: a myofibrillar network and transverse cytoskeletal connections are possible transmitters. *Journal of Cellular Physiology, 114,* 346–64.

Street, S. F., Sheridan, M. N., & Ramsey, R. W. (1966). Some effects of extreme shortening on frog skeletal muscle. *Medical College of Virginia Quarterly, 2,* 90–9.

Stull, J. T., Blumenthal, D. K., Miller, J. R., & DiSalvo, J. (1982). Regulation of myosin phosphorylation. *Journal of Molecular and Cellular Cardiology, 14* (Suppl. 3), 105–10.

Sugi, H., & Tsuchiya, T. (1981). Isotonic transients in frog muscle fibres following quick changes in load. *Journal of Physiology, London, 319,* 219–38.

Sugi, H., & Tsuchiya, T. (1988). Stiffness changes during enhancement and deficit of isometric force by slow length changes in frog skeletal muscle fibers. *Journal of Physiology, London, 407,* 215–29.

Sutko, J. L., Ito, K., & Kenyon, J. L. (1985). Ryanodine: a modifier of sarcoplasmic reticulum calcium release in striated muscle. *Federation Proceedings, 44,* 2984–8.

Sweeney, H. L., & Stull, J. T. (1986). Phosphorylation of myosin in permeabilized mammalian cardiac and skeletal muscle cells. *American Journal of Physiology, 250,* C657–60.

Sweeney, H. L., & Stull, J. T. (1990). Alteration of cross-bridge kinetics by myosin light chain phosphorylation in rabbit skeletal muscle: implications for regulation of actin–myosin interaction. *Proceedings of the National Academy of Sciences, U.S.A. 87,* 414–18.

Takeshima, H., Nishimura, S., Matsumoto, T., Ishida, H., Kangawa, K., Minamino, N., Matsuo, H., Ueda, M., Hanaoka, M., Hirose, T., & Numa, S. (1989). Primary structure and expression from complementary DNA of skeletal muscle ryanodine receptor. *Nature, London, 339,* 439–45.

Tanabe, T., Beam, K. G., Powell, J. A., & Numa, S. (1988). Restoration of excitation–contraction coupling and slow calcium current in dysgenic muscle by dihydropyridine receptor complementary DNA. *Nature, London, 336,* 134–9.

Tanabe, T., Takeshima, H., Mikami, A., Flockerzi, V., Takashahi, H., Kangawa, K., Kojima, M., Matsuo, H., Hirose, T., & Numa, S. (1987). Primary structure of the receptor for calcium channel blockers from skeletal muscle. *Nature, London, 328,* 313–8.

Taylor, S. R., & Rüdel, R. (1970). Striated muscle fibers: inactivation of contraction induced by shortening. *Science, 167,* 882–4.

Taylor, S. R., Rüdel, R., & Blinks, J. R. (1975). Calcium transients in amphibian muscle. *Federation Proceedings, 34,* 1379–81.

Ter Keurs, H. E. D. J., Iwazumi, T., & Pollack, G. H. (1978). The sarcomere length–tension relation in skeletal muscle. *Journal of General Physiology. 72,* 565–92.

Thames, M. D., Teichholz, L. E., & Podolsky, R. J. (1974). Ionic strength and the contraction kinetics of skinned muscle fibres. *Journal of General Physiology, 63,* 509–30.

Thomas, L. J. (1960). Ouabain contracture of frog heart: Ca^{45} movements and the effect of EDTA. *American Journal of Physiology, 199,* 146–50.

Thorens, S., & Endo, M. (1975). Calcium-induced calcium release and "depolarization" induced calcium release: their physiological significance. *Proceedings of the Japan Academy, 51,* 473–8.

Tobacman, L. S. (1987). Activation of actin–cardiac myosin subfragment-1 MgATPase rate by Ca^{2+} shows cooperativity intrinsic to the thin filament. *Biochemistry, 26,* 492–7.

Tobacman, L. S., & Sawyer, D. (1990). Calcium binds cooperatively to the regulatory sites of the cardiac thin filament. *Journal of Biological Chemistry, 265,* 931–9.

Toyoshima, Y. Y., Kron, S. J., McNally, E. M., Niebling, K. R., Toyoshima, C., & Spudich, J. A. (1987). Myosin subfragment-1 is sufficient to move actin filaments *in vitro. Nature, London, 328,* 536–9.

Trybus, K. M., & Taylor, E. W. (1980). Kinetic studies of the cooperative binding of subfragment 1 to regulated actin. *Proceedings of the National Academy of Sciences, USA, 77,* 7209–13.

Tsien, R. Y. (1983). Intracellular measurements of ion activities. *Annual Reviews of Biophysics and Bioengineering, 12,* 91–116.

Tsuchiya, T., & Edman, K. A. P. (1990). Mechanism of force enhancement after stretch in intact single muscle fibres of the frog. *Acta Physiologica Scandinavica, 140,* 23A.

Tsukita, S., & Yano, M. (1988). Instantaneous view of actomyosin structure in shortening muscle. In *Molecular Mechanism of Muscle Contraction,* ed. H. Sugi & G. H. Pollack, pp. 31–8. New York: Plenum Press.

Ueno, H., & Harrington, W. F. (1981). Conformational transition in the myosin hinge upon activation of muscle. *Proceedings of the National Academy of Sciences, USA, 78,* 6101–5.

Ueno, H., & Harrington, W. F. (1986). Local melting in the subfragment-2 region of myosin in activated muscle and its correlation with contractile force. *Journal of Molecular Biology, 190,* 69–82.

van der Laarse, W. J. Diegenbach, P. C., & Hemminga, M. A. (1986). Calcium-stimulated myofibrillar ATPase activity correlates with shortening velocity of muscle fibres in *Xenopus laevis. Histochemical Journal, 18,* 487–96.

Veratti, E. (1902). Richerche sulla fine struttura della fibra muscolare striata. *Memorie del R. Istituto Lombardo di Scienze e Lettere, 19,* 87–133.

Veratti, E. (1961). Investigations on the fine structure of a striated muscle fiber. *Journal of Biophysical and Biochemical Cytology, 10* (4), Suppl., 1–59. (Paper from 1902, translated by C. Bruni, H. S. Bennett, F. deKoven, & D. deKoven).

Vergara, J., & Delay, M. (1986). A transmission delay and the effect of temperature at the triadic junction of skeletal muscle. *Proceedings of the Royal Society [B], 229* 97–110.

Vergara, J. L., Rapoport, S. I., & Nassar-Gentina, V. (1977). Fatigue and post-tetanic potentiation in single muscle fibers of the frog. *American Journal of Physiology, 232,* C185–190.

Volpe, P., Gutweniger, H. E., & Montecucco, C. (1987). Photolabeling of the integral proteins of skeletal muscle sarcoplasmic reticulum: comparison of junctional and nonjunctional membrane fractions. *Archives of Biochemistry and Biophysics, 253,* 138–45.

Volpe, P., Krause, K. H., Hashimoto, S., Zorzato, F., Pozzan, T., Meldolesi, J., & Lew, D. P. (1988). Calciosome, a cytoplasmic organelle: the inositol 1,4,5-trisphosphate-sensitive Ca^{2+} store of nonmuscle cells? *Proceedings of the National Academy of Sciences, USA, 85,* 1091–5.

Wagenknecht, T., Grassucci, R., Frank, J., Saito, A., Inui, M., & Fleischer, S. (1989). Three-dimensional architecture of the calcium channel/foot structure of sarcoplasmic reticulum. *Nature, London, 338,* 167–70.

Wang, G. K., Brodwick, M. S., Eaton, D. C., & Strichartz, G. R. (1987). Inhibition of sodium currents by local anaesthetics in cloramine-T-treated squid axons. *Journal of General Physiology, 89,* 645–67.

Webb, M. R., Hibberd, M. G., Goldman, Y. E., & Trentham, D. R. (1986). Oxygen exchange between P_i in the medium and water during ATP hydrolysis mediated by skinned fibers from rabbit skeletal muscle. *Journal of Biological Chemistry, 261,* 15557–64.

Weber, A., Herz, R., & Reiss, I. (1963). On the mechanism of the relaxing effect of fragmented sarcoplasmic reticulum. *Journal of General Physiology, 46,* 679–702.

Weber, A., & Murray, J. M. (1973). Molecular control mechanism in muscle contraction. *Physiological Reviews, 53,* 612–73.

Weber, A., & Winicur, S. (1961). The role of calcium in the superprecipitation of actomyosin. *Journal of Biological Chemistry, 236,* 3198–202.

Weber, H. H. (1955). The link between metabolism and motility of cells and muscles. *Symposia of the Society for Experimental Biology. Fibrous Proteins and Their Biological Significance, 9,* 271–81.

Weisberg, A., Winegrad, S., Tucker, M., & McClellan, G. (1982). Histochemical detection of specific isozymes in rat ventricular cells. *Circulation research, 51,* 802–9.

Westerblad, H., & Lännergren, J. (1986). Force and membrane potential during and after fatiguing, intermittent tetanic stimulation of single *Xenopus* muscle fibres. *Acta Physiologica Scandinavica, 128,* 369–78.

Westerblad, H., & Lännergren, J. (1987). Tension restoration with caffeine in fatigued *Xenopus* muscle fibres of various types. *Acta Physiologica Scandinavica, 130,* 357–8.

Westerblad, H., & Lännergren, J. (1988). The relation between force and intracellular pH in fatigued, single *Xenopus* muscle fibres. *Acta Physiologica Scandinavica, 133,* 83–9.

Wilkie, D. R. (1981). Shortage of chemical fuel as a cause of fatigue: studies by nuclear magnetic resonance and bicycle ergometry. In *Human Muscle Fatigue: Physiological Mechanisms,* ed. R. Porter & J. Whelan, pp. 102–114. London: Pitman Medical.

Williams, D. L., Green, L. E., & Eisenberg, E. (1988). Cooperative turning on of myosin subfragment 1 adenosinetriphosphatase activity by the troponin–tropomyosin–actin complex. *Biochemistry, 27,* 6987–93.

Wilson, D. M., Smith, D. O., & Dempster, P. (1970). Length and tension hysteresis during sinusoidal and step function stimulation of arthropod muscle. *American Journal of Physiology, 218,* 916–22.

Winegrad, S. (1965). Autoradiographic studies of intracellular calcium in frog skeletal muscle. *Journal of General Physiology, 48,* 455–79.

Winegrad, S. (1971). Studies of cardiac muscle with a high permeability to calcium produced by treatment with ethylenediamine-tetraacetic acid. *Journal of General Physiology, 72,* 737–64.

Winegrad, S. (1984). Regulation of cardiac contractile proteins: correlations between physiology and biochemistry. *Circulation Research, 55,* 565–74.

Winegrad, S., & Shanes, A. M. (1961). Ca flux and contractility in guinea pig atria. *Journal of General Physiology, 45,* 371–94.

Winegrad, S., & Weisberg, A. (1987). Isozyme specific modification of myosin ATPase by cAMP in rat heart. *Circulation Research, 60,* 384–92.

Winegrad, S., Weisberg, A., Lin, L.-E., & McClellan, G. (1986). Adrenergic regulation of myosin adenosine triphosphatase activity. *Circulation Research, 58,* 83–95.

Yagi, S., & Endo, M. (1980). Effects of dibucaine on skinned skeletal muscle fibers. An example of multiple actions of a drug on a single subcellular structure. *Biomedical Research, 1,* 269–72.

Yanagida, T., Arata, T., & Oosawa, F. (1985). Sliding distance of actin filament induced by a myosin crossbridge during one ATP hydrolysis cycle. *Nature, London, 316,* 366–9.

Yates, L. D. (1985). Calcium activation of fibers. *Reciprocal Meat Conference Proceedings, 38,* 9–25.

Yu, L. C., Hartt, J. E., & Podolsky, R. J. (1979). Equatorial x-ray intensities and isometric force levels in frog sartorius muscle. *Journal of Molecular Biology, 132,* 53–67.

Yue, D. T., Marban, E., & Wier, W. G. (1986). Relationship between force and intracellular [Ca^{2+}] in tetanized mammalian heart muscle. *Journal of General Physiology, 87,* 223–42.

Zot, A. S., & Potter, J. D. (1987). Structural aspects of troponin–tropomyosin regulation of skeletal muscle contraction. *Annual Review of Biophysics and Biophysical Chemistry, 16,* 535–59.

Index

Printed in the United States
By Bookmasters